LAND MINES IN ANGOLA

An Africa Watch Report

Human Rights Watch

New York • Washington • Los Angeles • London

Copyright © January 1993 by Human Rights Watch.
All rights reserved.
Printed in the United States of America.

Library of Congress Catalog Card No.: 93-77288
ISBN 1-56432-091-X

Africa Watch

Africa Watch was established in May 1988 to monitor and promote observance of internationally recognized human rights in Africa. The chair of Africa Watch is William Carmichael. Alice Brown is the vice chair. Janet Fleischman and Karen Sorensen are research associates. Urmi Shah and Ben Penglase are associates.

Human Rights Watch

Human Rights Watch is composed of Africa Watch, Americas Watch, Asia Watch, Helsinki Watch, Middle East Watch and the Fund for Free Expression.

The executive committee is comprised of Robert L. Bernstein, chair; Adrian W. DeWind, vice chair; Roland Algrant, Lisa Anderson, Peter Bell, Alice Brown, William Carmichael, Dorothy Cullman, Irene Diamond, Jonathan Fanton, Jack Greenberg, Alice H. Henkin, Stephen Kass, Marina Pinto Kaufman, Jeri Laber, Aryeh Neier, Bruce Rabb, Harriet Rabb, Kenneth Roth, Orville Schell, Gary Sick and Robert Wedgeworth.

The staff includes Aryeh Neier, executive director; Kenneth Roth, deputy director; Holly J. Burkhalter, Washington Director; Gara LaMarche, associate director; Ellen Lutz, California director; Susan Osnos, press director; Jemera Rone, counsel; Kenneth Anderson, Arms Project director; Joanna Weschler, Prison Project director; Dorothy Q. Thomas, Women's Rights Project director; and Allyson Collins, research associate.

Executive Directors

Africa Watch	Americas Watch	Asia Watch
	Juan E. Méndez	Sidney Jones
Helsinki Watch	Middle East Watch	Fund for Free Expression
Jeri Laber	Andrew Whitley	Gara LaMarche

Addresses for Human Rights Watch:

485 Fifth Avenue
New York, NY 10017
Tel: (212) 972-8400
Fax: (212) 972-0905
Email: hrwatchnyc@igc.qpc.org

1522 K Street, NW, Suite 910
Washington, DC 20005
Tel: (202) 371-6592
Fax: (202) 371-0124
Email: hrwatchdc@igc.apc.org

10951 West Pico Blvd., #203
Los Angeles, CA 90064
Tel: (213) 475-3070
Fax: (213) 475-5613
Email: hrwatchla@igc.qpc.org

90 Borough High Street
London SE1 1LL, UK
Tel: (071) 378-8008
Fax: (071) 378-8029
Email: africawatch@gn.apc.org

For information on publications please contact our New York office.

CONTENTS

INTRODUCTION AND SUMMARY 1

1. HISTORY OF THE CONFLICTS IN ANGOLA 4
   Colonial Rule 4
   Independence and Civil War 5
   The Peace Process 8
   The Impact of the War 9
   The Future 11

2. TECHNICAL ASSESSMENT OF MINELAYING 13
   Mine Types and Sources 13
      Anti-personnel mines 13
      Directional devices—remote or trip
      initiation 17
      Anti-tank devices 18
   Dissemination Strategies Employed 19
      Route denial 19
      Ambush 19
      Bridgehead Mining 20
      Defensive Mining of Key Structures and
      Facilities 20
      Random Dissemination 21
   Use of Improved Explosive Devices and
   Booby-Traps 21
   Land Mine Records 22
   Assessments of the Total Land Mine Threat ... 23
   United States Involvement 25

3. THE HUMAN COST 26
   Who are the Victims 26
   Where the Mines are Planted 28
   Who Laid the Mines 34
   Knowledge about Minefields and Warnings
   Given 35
   Emergency Care for the Injured 36
   Medical Care and Rehabilitation 36

  Social Rehabilitation 38

4. THE SOCIAL AND ECONOMIC IMPACT 41
  Repatriation 42
  Moxico and the Lundas 42
  Zaire, Uíge and Malanje 44
  Overall Impact 47

5. MINE CLEARANCE INITIATIVES 49
  FAPLA/FALA Teams 49
  SADF Involvement 51
  British Army Initiatives 52
  United States Involvement 54
  Equator Bank, USA, Initiative 55
  The Cap Anamur Initiative 56
  Conclusion 57

6. LAND MINES IN INTERNATIONAL LAW 58
  The Basic Rule: Protecting Civilians and Civilian
  Objects 59
  Prohibition of Disproportionate Attacks 61
  Prohibition Against Starvation of the Civilian
  Population 62
  Recording Requirement 64

CONCLUSIONS AND RECOMMENDATIONS 66
  Conclusions 66
  Recommendations 68
  I. General 68
  II. To the Angolan Government 68
  III. To FAPLA, FALA/UNITA, and the Cuban
  Armed Forces 69
  IV. To the United Nations, Western Donors and
  Former Eastern Bloc Countries 69

INTRODUCTION AND SUMMARY

To the homes, to our camps, to the beaches, to our fields, we shall return.

Agostinho Neto, 1960

When Agostinho Neto, the first leader of Angola's liberation struggle, spoke those words, he cannot have imagined that more than three decades later, hundreds of thousands of Angolans would still be waiting to return to their homes, camps, beaches and fields. Even during the relative peace that prevailed between the signing of the ceasefire in May 1991 and the elections in September 1992, much of the country remained uninhabitable or dangerous. One of the main reasons for this is the vast number of land mines in Angola. These mines have claimed tens of thousands of victims.

At the time of writing, the future of Angola looks bleak. A return to fighting, following the disputed election results of September 1992, has claimed thousands of lives. No doubt, the widespread and indiscriminate use of anti-personnel land mines will continue to be a significant feature of the war, as it has in the past. This report demonstrates that the use of land mines by both sides represents a gross violation of basic human rights. Both parties to the conflict deserve international opprobrium if they return to their former tactics of land mine usage and do not clear mines laid in the past.

If peace returns, civilian casualties due to land mines will continue. One victim interviewed by Africa Watch is A. da S., who stepped on a mine outside a bar in his home town of Camanongue in Moxico Province, the day after returning home just weeks after the ceasefire was signed in May 1991. If normality returns, and Angolans begin to reclaim the use of their land, casualties will mount. Mines have rendered large areas of arable land and pasture, many roads, bridges, riverbanks and villages, and some important economic installations, off-limits to people. This report documents how this tragedy came about and its devastating consequences for the Angolan people. It also documents what efforts are being made to undo some of the damage. The value of these efforts is extremely uneven, and a more concerted attempt to eradicate land mines in Angola will be needed in the future.

Chapter one consists of a brief history of Angola and the wars that have ravaged the country for the last thirty years.

Chapter two is a technical assessment of minelaying in Angola. It examines the makes and types of mines that have been used, and the methods of their use. Angola itself does not manufacture mines; all the devices found in the country have been supplied from abroad by manufacturers, governments and arms dealers who are thereby complicit in the maiming and death of tens of thousands of Angolan civilians. Africa Watch has confirmed that thirty-seven types of mine have been used in Angola, and the number is probably greater. The total number of mines in Angolan soil cannot be known, but runs into the hundreds of thousands or millions. Mines have been laid for a variety of military purposes, such as protecting key installations and denying the use of roads and bridges to the enemy. Few of these minefields have been recorded or marked; there have rarely been any attempts to protect civilians from the dangers they pose. Perhaps the most common use of mines has been their random dissemination in and around villages. While there may have been a slender military rationale for this use, its main impact has been to render paths, fields and villages unusable to civilians except at great personal danger, thereby terrorizing the community. This use of land mines is illegal.

Chapter three examines the human impact of the land mines. Angola suffers from one of the highest per head ratios of land mine victims in the world. At least 15,000 Angolans are amputees as a consequence of stepping on land mines; about half of these are soldiers and half civilians. There are more disabled men than women on account of land mines, but this may reflect higher fatality rates for women during and after land mine accidents. Children are also victims. Civilians are injured in their fields, on paths, roads, riverbanks, and inside built-up areas. The medical care and physical and social rehabilitation of these people is a challenge and a burden to Angola. Evacuation and medical facilities are inadequate. At least 5,000 prostheses will be needed each year for the foreseeable future for the amputees, far more than are currently manufactured.

Chapter four looks at the wider social, economic and political impact of the mines. Land mines represent a formidable obstacle to commerce and free movement, to economic reconstruction, and to the effective delivery of relief and other forms of aid. The chapter looks particularly at how mines are preventing the rapid and safe return of refugees.

Chapter five provides an account of current initiatives to clear land mines. The major program has consisted of joint clearance efforts by the two armies. Major roads, the railways, key economic installations and towns have been cleared or are in the process of clearance. However, the teams lack much basic equipment and have no effective central coordination. In addition, there are a number of foreign governments, private companies and humanitarian organizations involved in mine clearance operations. Some of these programs are seriously flawed, for example the British "training" program. Other governments and companies are notable by their absence or small scale of involvement, for example the United States.

Chapter six consists in an overview of the legal regime ostensibly governing the use of land mines, specifically the 1981 Land Mines Protocol. It is evident that the great majority of land mines in Angola have been deployed in flagrant disregard for the provisions of the Protocol. In fact, the Land Mines Protocol has proved wholly irrelevant to the conflict in Angola, as it has been shown to be unworkable elsewhere in the world. Africa Watch concludes that only a complete ban on the use of anti-personnel land mines can remove the unreasonable danger that they pose to civilians.

This report is based upon a visit to Angola in April-May 1992 by Rae McGrath, director of the Mines Advisory Group and a consultant to Africa Watch. Additional material has been taken from visits to Angola in November-December 1990 by Jemera Rone, Counsel to Human Rights Watch, and September-October 1992 by Alex Vines, a consultant to Africa Watch, and from material collected in the U.S. by Ben Penglase, an Associate of Africa Watch. The chapter on the legal regime was written by Jemera Rone. The report was edited by Alex de Waal, consultant to Africa Watch.

This report was made possible by a grant to Africa Watch from Oxfam (UK), whose assistance is gratefully acknowledged.

1. HISTORY OF THE CONFLICTS IN ANGOLA

Angola has rarely known peace, and has never experienced democratically accountable government, respect for human rights, or prosperity. The period between 1975 and 1991 saw a particularly brutal civil war between the Movimento Popular de Libertação de Angola (MPLA), in government, and the União Nacional para a Independência Total de Angola (UNITA). This came to an end with a peace accord signed in Lisbon on May 31, 1991. Following this, there was a period of peace until the country's first multi-party elections were held in September 1992. The results of the Presidential contest were inconclusive, but MPLA leader President Eduardo Dos Santos had a significant lead over UNITA leader Jonas Savimbi. At the time of writing, fighting has resumed, and the future of the country looks grim.

Should peace be established, Angola will face a huge struggle to heal the wounds caused by the long and bitter civil war. This process will be helped by Angola's vast natural wealth. In addition to large reserves of fertile land, Angola possesses an abundance of diamonds and, above all, oil.

Angola's population is not known for sure, but is currently estimated at between nine and ten million people. This consists of various ethnic groups. The Ovimbundu are the largest single group, forming about 37 percent of the population; the Mbundu form 23 percent and the Kongo 14 percent. A number of other peoples make up the remainder.

Colonial Rule

Angola was colonized by the Portuguese, initially to secure the coastline and to obtain slaves for their possessions in Brazil. In the late nineteenth century Portugal began to establish cotton plantations for the benefit of its domestic textile industry, managed by Portuguese settlers. Angola was also the site of an important coffee industry, also run by the Portuguese settlers.

Unlike the British, French and Belgian colonial rulers, the Portuguese government decided against granting independence to its African colonies in the 1950s and '60s. This led to long and bloody independence struggles in each of Portugal's colonies. The first shots in the Angolan liberation war were fired in January 1961. Three nationalist movements were formed. The first was the MPLA, founded in 1956, and drawing most of its support from the Mbundu ethnic group around

Luanda, Angola's capital, and mixed-race Angolans. From early days, the MPLA was a leftist organization and obtained support from the USSR and Cuba. It was led by Dr. Agostinho Neto until his death in 1979. The Frente Nacional de Libertação de Angola (FNLA) was originally founded in 1957, drawing its support chiefly from the Kongo of the north. Its leader was Holden Roberto, a friend of President Mobutu Sese Seko of Zaire. It had a nationalist ideology and received support from Zaire. In 1966, the foreign minister of the FNLA, Jonas Savimbi, denounced Roberto as an agent of "American Imperialism" and broke away to form UNITA. Savimbi drew most of his support from the Ovimbundu. UNITA obtained outside assistance from Zambia and later South Africa, but for several periods had no significant outside support at all. Savimbi espoused a variety of populist and nationalist ideologies, but the dominating factor has been a personality cult built around the leader himself.

One of the tragedies of modern Angolan history has been the inability of the three movements to form a united front. They each fought the Portuguese fiercely, but also attacked each other and did not come close to achieving a military victory.

Independence and Civil War

Following a military coup in Portugal in April 1974, the colonial government precipitously announced its withdrawal from Angola. Along with the colonial government, 90 percent of the 335,000 Portuguese settlers left. As the Portuguese had dominated all employment that required even minimal education, this dealt a devastating economic blow. Much of the basic functioning of government came to a halt. The resulting economic crisis was a severe blow, made much worse by the continuation of the war. In the following ten years, the war was to do damage worth an estimated $17.6 billion to the economy. Diamond mining and coffee production collapsed. Only oil, most of it produced in the Cabinda enclave to the north of the main part of the country, continued unchecked, providing a vital economic base for the government. Oil provides 90 percent of the government's foreign exchange.

In January 1975, the three movements signed the Alvor Accord agreeing to a joint interim government and an integrated national army. However, as the date for military integration neared, the agreement broke down. By mid-1975, the fronts were at war. The superpowers and regional powers rushed to involve themselves. The United States had already granted covert aid to the FNLA in January 1975. The USSR and Cuba supported the MPLA, which was able to seize control of Luanda, but little else. South Africa invaded Angola in support of UNITA and Zaire invaded in support of the FNLA; by October, it looked as though Luanda would be captured before the official date of independence, November 11. However, a massive Soviet airlift of military equipment and Cuban troops reversed the military tide. Zaire abandoned its invasion force and the South Africans withdrew. The revelation of South African backing for UNITA and FNLA was disastrous for the reputation of the two movements in Africa, and the MPLA was able to form a one-party socialist government that obtained widespread diplomatic recognition. The US, however, refused to recognize the MPLA government.

In retaliation for Zairean support for the FNLA, Angola backed Katangese separatist forces in their opposition to Mobutu. The Zairean rebels mounted an unsuccessful invasion in 1978. Later that year, Angola and Zaire signed an agreement to stop harboring each others' military opponents. This led directly to the effective military demise of the FNLA—though it has remained as a small political movement. However, in the early 1980s, UNITA pushed northwards into areas formerly controlled by the FNLA, and began collaboration with the Zairean government.

Despite the success of the Forças Armadas Popular para a Libertação de Angola (FAPLA, the Angolan army) in 1976, in the south and east, UNITA's resistance continued. The South African Defence Force (SADF) intermittently operated inside Angola in support of UNITA. In 1976, the SADF also formed 32 or Buffalo Battalion, from Angolan refugees, to fight inside Angola. The largest South African incursions were three invasions in 1981-3. South African involvement was in part in retaliation for the MPLA government's support of the South West African People's Organization (SWAPO) in its guerrilla war against South African-occupied Namibia. South African incursions were often aimed at SWAPO bases inside Angolan territory.

In late 1983, the UN Security Council demanded that South Africa withdraw from Angola. Shortly afterwards, Angola and South Africa signed the Lusaka Accords, under which South Africa agreed to withdraw if Angola ceased support for SWAPO. However, South African withdrawal was extremely slow, and was reversed in 1985 when another invasion was launched, in support of UNITA which was facing defeat against a full-scale attack by FAPLA with Cuban support. The government clearly believed that if South African support for UNITA was withdrawn, it would be able to achieve a military solution to the conflict.

US covert assistance to UNITA, which had been suspended by the Clark Amendment in 1976, was restarted after the repeal of the Clark Amendment in 1985. The US began to supply significant covert funding to UNITA in that year, and starting in 1986 provided Stinger anti-aircraft missiles.

The war intensified again in 1987, with major battles in the south of the country. A major South African military presence remained, and in November 1987 President P. W. Botha actually visited South African troops inside Angola. The SADF-UNITA plan was the capture of the besieged FAPLA-Cuban forward base of Cuito Cuanavale. Despite massive conventional assaults, the siege was lifted, and in March 1988 the South Africans decided that they could not win. This also marked the failure of UNITA to move from guerrilla warfare into mobile warfare, and the end of any prospects of an outright military victory over FAPLA.

The SADF-UNITA defeat at Cuito Cuanavale marked the beginning of negotiations to end the conflict. The South Africans began talks with their adversaries at Brazzaville (Congo) and London. An agreement was reached by the Angolan government, Cuba and South Africa, mediated by the US and USSR, whereby South Africa and Cuba were to withdraw their troops and Namibia was to be granted its independence. These elements of the agreement were implemented according to plan. However, despite some South African pressure, UNITA was not included in the talks, and no mention was made of the internal conflict between the MPLA and UNITA.

Throughout this period, US covert support for UNITA continued, as did massive Soviet military assistance to the government. Throughout the 1980s, the government spent approximately $1 billion per year on its military, about 20 percent of gross national product. This level of investment in war was only matched in Africa by the former Ethiopian government. It discouraged the government from seeking to negotiate a solution to the conflict.

Following the South African withdrawal, UNITA appeared weak. Its southern headquarters at Jamba was isolated and vulnerable, and it made preparations to launch a new guerrilla offensive in the north, operating across the border from Zaire. This strategy was designed with US assistance; in 1988 the US and Zairean armies conducted joint military maneuvers "with the long term aim of providing UNITA with logistical support for a northern base to be sited close to the town of

Quimbele."[1] (In 1990, UNITA was forced to withdraw in the face of a major FAPLA offensive in the south, but Jamba remained safe; meanwhile intensified UNITA operations began in the north.)

The Peace Process

Also in 1988, the USSR signalled that it was no longer prepared to continue arming the Luanda government ad infinitum; it would be necessary for the MPLA to negotiate an end to the war. In January 1989, President Dos Santos made the first peace offer to UNITA. This was a radical break with his former warlike stance, but still fell short of recognizing that UNITA was a political as well as a military force. This led to a peace process brokered by eight African nations and a meeting in Gbadolite, Zaire, on June 22, at which Dos Santos and Savimbi shook hands and agreed on an immediate ceasefire. However, the details of the agreement, which remained secret, were disputed by Savimbi shortly afterwards; he denied that he had agreed to go into exile while UNITA was integrated into the MPLA government. By July, the war had resumed, with UNITA launching major offensives.

In August, the African leaders met again in Harare, Zimbabwe, to try to salvage the peace agreement. President Mobutu of Zaire, the chief mediator in the Gbadolite agreement, was heavily criticized for his hasty and vague manner of negotiation, and was instructed to pressure Savimbi to accept the deal. Savimbi angrily rejected this and accused Mobutu of not being a neutral mediator—an ironic charge, as Zaire was now the main base for UNITA's supply, and Dos Santos was also unhappy with Mobutu's role. Retaliating to this charge, Mobutu briefly suspended the clandestine supply of US arms to UNITA through Zaire.

The following eighteen months were to see, simultaneously, the most sustained efforts to achieve a peaceful settlement, and some of the fiercest fighting of the entire war. In September and December 1989, FAPLA launched major offensives aimed at capturing the UNITA-held town of Mavinga in Cuando Cubango Province. UNITA was pushed back, and in February 1990, the town fell. The government then launched air strikes on the UNITA headquarters of Jamba, reportedly injuring Savimbi himself.

---

[1] *Africa Confidential*, May 27, 1988.

In October 1989, President Bush met with Savimbi and pressured him to resume peace negotiations. However, covert military assistance was continued. In March 1990, Secretary of State James Baker and Soviet Foreign Minister Eduard Shevardnadze met at the Namibian independence ceremony and agreed to cooperate in the search for peace. In April, the Portuguese government was named as mediator.

Five sets of inconclusive talks were held over the next eight months, deadlocked over UNITA's demand for formal recognition as a political party by the MPLA, the question of the formation of an integrated national army, and the timing of elections. Under pressure from the USSR, in December 1990, the MPLA announced the creation of a multi-party system and other far-reaching political and economic reforms. This broke through the diplomatic impasse and raised hopes for a ceasefire, and in January both sides agreed to a peace accord.

While the final rounds of the peace talks continued in early 1991, both sides mounted major military actions to try to secure their positions before the announcement of a ceasefire. There was heavy fighting in April as UNITA tried, unsuccessfully, to seize Luena, provincial capital of Moxico. Hundreds of civilians and soldiers were reportedly killed.

In Lisbon on May 31, 1991, President Eduardo dos Santos and Jonas Savimbi signed the peace accord that formally brought to an end the civil war. The agreement specified that a civilian government should be established based on the result of free multi-party elections, monitored by the UN. The two armies—FAPLA and UNITA's army Forças Armadas para a Libertação de Angola (FALA)—were to be integrated, with the majority of the estimated 300,000 soldiers being demobilized. A joint military commission was established to organize this. One of its tasks was to coordinate the clearance of land mines.

The Impact of the War

During the war, foreign powers have poured weaponry into Angola. The USSR supplied billions of dollars worth of military equipment to Angola, including 500 battle tanks and over 150 combat aircraft. The Cuban military presence peaked at 50,000 troops in 1988, and did not drop below 25,000 until well after the gradual withdrawal began in April 1989. The South African military effort was estimated to cost $2 billion in 1988 alone. The full extent of US covert assistance to UNITA has never been disclosed, but it certainly ran in the tens of millions of dollars in the late 1980s.

The war witnessed widespread human rights abuses by both sides.[2] There was much deliberate killing and wounding of civilians. Civilian properties were frequently looted or destroyed. Whole villages have been burned. Foodstuffs and other items indispensable for the survival of civilians, such as cattle and plough oxen, were taken. UNITA took many civilians, including children, by force to serve in its armed forces or to be auxiliaries. FAPLA also recruited boys and men by force.

The war was fought in a manner that reduced much of Angola's population to a state of famine. There were no recognized front lines, and fighting raged backwards and forwards over large areas of the country. As a result, a very large proportion of the population was directly affected by the war, and an even larger number of people lived with the pervasive fear that fighting could come to their locality at any time. The widespread use of land mines, especially on roads and paths, was a crucial factor in creating famine. The threat of land mines prevented free movement of people and commerce, and proved a serious obstacle to relief efforts.

During 1990, serious food shortages threatened much of the country. According to estimates by the US Office of Foreign Disaster Assistance, up to 10,000 people died in the first four months of the year. In September, the United Nations estimated that about 1.9 million Angolans in nine central and southern provinces faced famine. About three quarters of those at risk were in areas made inaccessible for relief. About 1.2 million people were in the central Planalto of Huambo and Bíe provinces and the neighboring areas. This, the most fertile and densely populated part of Angola, was the center of UNITA's war effort. UNITA aimed to destabilize the government by preventing it from exercising any form of authority in these provinces. This strategy, together with the shifting battle lines, meant that the delivery of relief to the Planalto by establishing tranquil zones or safe passage agreements would be possible only if UNITA dramatically revised its military strategy.

In 1990, about 150,000 people were estimated to be at serious risk of famine in the UNITA-controlled southeastern provinces of Moxico and Cuando Cubango, with a much higher number suffering less extreme deprivation. Here, the chronic nature of the conflict had worn down

---

[2] *Angola: Violations of the Laws of War by both Sides*, Africa Watch, New York and London, 1989, and "Angola: Civilians devastated by 15-year war," *News from Africa Watch*, February 5, 1991.

much of the population to the brink of complete destitution. Because the area was securely controlled by UNITA, cross-border relief from Zambia or Namibia was a possibility (and indeed was provided by at least one humanitarian organization). However, consent for a larger relief operation by the UN or International Committee of the Red Cross (ICRC) depended upon government permission. The Angolan government was emphatic that it would not compromise its sovereignty by allowing cross-border relief efforts into these areas, without a general ceasefire. It rejected UNITA plans for corridors from Namibia, claiming that they would be used to smuggle arms.

After diplomatic pressure, the Angolan government began to consider proposals for neutral relief corridors in July 1990. US and UN missions to Luanda began to discuss details of how such a scheme would work. After prolonged negotiations between the US State Department and the MPLA government, a plan for relief corridors was announced on September 24. Five "peace corridors" were envisaged, including two cross-border from Namibia. The UN coordinated the deliveries, and the ICRC and private voluntary agencies participated. The first convoy moved on November 2.

Both sides were deeply mutually suspicious and ready to use famine relief as a weapon. The program met repeated delays from both sides, and on December 21 the government halted the relief corridors, alleging that UNITA had violated the agreement by destroying a bridge. The UN immediately suspended its operations. No more convoys moved along the peace corridors until March 1991.

The Future

Under United Nations supervision, Angola held its first multi-party elections on September 29 and 30, 1992. In the 220-seat assembly, the MPLA won 129 seats and UNITA won 70. In the Presidential election, Jose Eduardo dos Santos won 49.6 percent of the votes and Jonas Savimbi received 40.1 percent. Nine other candidates, including FNLA leader Holden Roberto, shared the remainder. Because dos Santos had failed to win an outright majority, a run-off between him Savimbi was scheduled to occur within thirty days.

This did not occur. Savimbi denounced the election as rigged. International observers have said that although there were some irregularities, the elections were generally conducted fairly.

As soon as the projected election results became public, Savimbi withdrew his forces from the nascent joint army, while the MPLA also began to make preparations for a military conflict. At this writing, UNITA now occupies large swathes or rural Angola and has captured key towns such as Uíge. Civilian militias loyal to the MPLA have engaged in round-ups of suspected UNITA supporters in Luanda and other otwns, leading to numerous summary executions. Mixed-race Angolans, mostly identified with the MPLA, have been targets for UNITA attacks. Thousands have already died in the four months since elections. Millions more reportedly face starvation in the next few months if the fighting continues. [3]

At the time of writing, negotiations are just beginning in Addis Ababa, and their outcome is uncertain. UN Secretary General Boutros Boutros-Gali has recommended an April 30, 1993 deadline for the warring parties to make peace, and has said that the Security Council should withdraw its peace-keeping forces it the deadline is not met. He has also recommended reducing the UN force to about 60 observers, to be concentrated in Luanda.

---

[3] Associated Press, January 28, 1993/

2. TECHNICAL ASSESSMENT OF MINELAYING

This chapter examines the types of land mine used in Angola, their origin, and the strategies used by the parties to the conflict to disseminate them. While minelaying was practiced from 1961 onwards until 1991, the great majority of mines were laid in the period between 1975 and 1988. No detailed or reliable records exist which could form the basis of finding out which mines were obtained by the warring armies, in what numbers, and where and how they were used. The information contained in this report has been obtained entirely from Africa Watch's own investigations, and information gathered in a piecemeal fashion by those responsible for mine clearance activities.

Mine Types and Sources

Africa Watch has confirmed that the following thirty-seven types of mine have been deployed in Angola. This is based on physical inspection of the mines themselves or detailed descriptions of them. It is likely that some additional types have been used, but Africa Watch has not been able to obtain reliable evidence for these.

A. Anti-personnel mines

1. *M409 anti-personnel blast mine.*

    This mine is initiated by 8-30kg pressure. It has a very low metallic content (limited to one gram in fuse and aluminum powder in explosive charge) and is thus difficult to detect.
    Manufactured by PRB of Belgium.

2 & 3. *72 and 72b anti-personnel mines*

    These small plastic mines have been the subject of a great amount of misinformation and have attained almost cult status in some countries. There are actually two different devices that share the same outer casing. The 72 is not detectable by mine detectors, and the 72b is.
    The 72 (sometimes called 72a) is a small light-green colored plastic mine with a fabric-covered pressure plate

containing 34 grams of explosive and requiring between 3 and 7kg of pressure for initiation. Due to the extremely low metallic content the mine is virtually undetectable using most detectors.

The 72b is identical once deployed (the arming pin ring is shaped differently but this is discarded when the mine is deployed) but is fitted with a secondary tilt mechanism which initiates the mine when it is tilted through more than approximately fifteen degrees. This mine can also be initiated by 3-7kg of pressure. The 72b is easily detected by a detector due to the high metallic content of the tilt mechanism but presents a high risk when using manual probing techniques.

Manufactured in China.

4. *Valmara VS-69 bounding anti-personnel mine*

This bounding mine is filled with either 650 6mm steel ballbearings or 1,200 4mm steel cubes which act as shrapnel. It can be initiated by either 10kg of direct pressure on the fuse prongs or 6kg exerted on a tripwire. Upon initiation, the mine is fired to approximately 1.2 meters vertically on a tether wire before exploding; it has a killing zone of 27 meters throughout an arc of 360 degrees. In other words, it explodes at the height of a man's chest, and has the power to rip out the heart of anyone standing within one hundred feet.

Manufactured by Valsella of Italy.

5. *VS Mk-2 scatterable anti-personnel mine*

A small low-metallic content plastic mine designed to be scattered on the ground or buried. It requires 10kg pressure to initiate an explosion.

Manufactured by Valsella of Italy.

6. *USK*

No reliable information has been obtained about the design or operation of this mine.

Manufactured in South Africa.

7 & 8.   *M16 A1 and M16 A2 bounding anti-personnel mine*

The M16 series of bounding anti-personnel mines employ the M605 fuse and can be initiated by 3.6-20kg direct pressure or between 1.6-3.8kg pull on a tripwire. The mine is propelled vertically to a height of one meter when initiated. The main charge of TNT then disperses metal fragmentation through 360 degrees with an effective range of approximately thirty meters. Like the Valmara VS-69, it explodes at waist or chest height and rips apart the torso of anyone within one hundred feet.

Manufactured in the USA.

9.   *M14 anti-personnel mine*

The M14 is a small plastic blast mine, only 40mm high by 56mm diameter. It requires 9-16kg of direct pressure to initiate an explosion. Metallic content is limited to the tip of the firing pin, making the mine difficult to detect using mine detectors.

10.   *PMN blast anti-personnel mine*

The PMN is a very common mine, undoubtedly the cause of a large percentage of mine-related deaths and amputations throughout the world. Although easily detected, this comparatively large device (56mm high by 112mm diameter) requires as little as 0.25kg of direct pressure to initiate an explosion. It contains 240 grams of TNT.

Manufactured by Soviet State Arsenals.

11.   *PMN2 blast anti-personnel mine*

A high technology development of the PMN requiring an initiation pressure of approximately 5kg.

Manufactured in the former USSR.

12.   *PMD-6 blast anti-personnel mine*

This mine employs a wooden box body with a block of cast TNT that is initiated when 1-10kg of downward pressure on the box forces the pin out of a MUV-2 fuse. The design has been widely copied. After having been buried for some time, this mine becomes unstable and finally ineffective when the wood rots. There is a high metallic content in the fuse, aiding detection.
Manufactured by Soviet State Arsenals.

13 & 14.   *POMZ-2 and POMZ-2M fragmentation anti-personnel mines*

Both these mines consist of a cast iron fragmentation casing mounted on a wooden stake. The casing contains a 75 gram charge of TNT and a fuse (normally an MUV fuse) which protrudes from the top of the casing. A tripwire is connected to a striker-retaining pin in the fuse and a pull of approximately 1kg on the tripwire will release the striker and initiate an explosion. The POMZ-2 has six rows of fragmentation while the later POMZ-2M has only five. Both mines have an effective killing range of up to 25 meters.
Manufactured by Soviet State Arsenals.

15.   *PP-Mi-SR bounding anti-personnel mine*

A metallic cased bounding mine which on initiation is projected to a tethered height of one meter before detonation. The casing of the mine acts as fragmentation. It is initiated by a tripwire and requires a pull of between 3-8kg depending on the fuse type fitted.
Manufactured in Czechoslovakia.

16.   *DM-11 anti-personnel mine*

The DM-11 is 37mm high by 81mm diameter with a 114 gram RDX/TNT main charge. It requires direct pressure of between 5-10kg for initiation.

Manufactured by DIEHL Ordnance Division, Rothenbach, (West) Germany.

17. *DM-31 mine*

No reliable information has been obtained about this mine, also manufactured in Germany.

18. *MIN-25-ANOS anti-personnel mine*

No reliable information has been obtained about this mine.

B.   Directional devices—remote or trip initiation

1. *M18A1 directional ambush mine ("Claymore")*

This mine consists of a curved fiber-glass casing containing 700 steel balls mounted in front of 682 grams of C4 plastic explosive. The mine can be fired remotely or may be tripwire initiated. (When fired by hand and aimed at a military objective, it is not indiscriminate.) It has a frontal killing area of fifty meters, to a height of two meters, through an arc of sixty degrees (when it is cited at ground level). There is a fifteen meter danger zone to the rear of the mine.
Manufactured by Morton Thiokol, Shreveport, Louisiana, USA, and also widely copied and produced under license in different countries.

2. *MON-50 directional ambush mine*

The MON-50 is a virtually identical Soviet derivative of the Claymore.
Manufactured by Soviet State Arsenals.

3. *MON-100 directional ambush mine*

The MON-100 is a larger version of the MON-50. The casing is cylindrical with a face diameter of 220mm. The mine contains 450 pieces of steel fragmentation mounted in 5kg of plastic explosive. The killing area is reported to be 100 meters. Manufactured by Soviet State Arsenals.

C. **Anti-tank devices**

Africa Watch recorded seventeen types of anti-tank mines in Angola, supplied from a wide range of countries.

Anti-tank mines are generally designed to incapacitate tanks, usually by causing damage to the tracks, final drive or idlers, although some are designed to pierce the armor and kill the crew by secondary fragmentation.

The pressure required to initiate an anti-tank device varies from 60-500kg depending on the make and design. Humans, animals and light vehicles may therefore pass over them safely. Although these mines present a lesser risk to civilians than do anti-personnel mines, the instances of cars and trucks (especially when heavily loaded) and their passengers being blown up by anti-tank mines are still common enough to be of serious concern. In addition to the initiation of anti-tank mines by normal direct pressure, they may also be initiated by lesser pressure when there is a fault in the mine or when an anti-personnel mine is laid on top of the anti-tank mine, or another secondary means of initiation is used.

Africa Watch recorded the following types of anti-tank mine in Angola.

1.		M6	Manufactured in USA
2.		M7	USA
3.		M15	USA
4.		M19	USA
5.		MK7	Great Britain
6.		TMA2	Yugoslavia
7.		TMA3	Yugoslavia
8.		TMA4	Yugoslavia
9.		TMA5	Yugoslavia
10.		PT-MI-BA3	Yugoslavia
11.		TM46	USSR
12.		TMN46	USSR
13.		TM57	USSR

14.	TM62	USSR
15.	T72	China
16.	No.8	South Africa
17.	MK3	South Africa

Dissemination Strategies Employed

As with all major military conflicts during the last thirty years, combatant forces have used land mines in various roles. In Angola these can be classified as follows:

A.   Route denial

This is a strategy employing primarily anti-tank and anti-vehicle devices in order to deny the use of roads and tracks to the opposing forces. In some cases, anti-tank mines are "protected" by anti-personnel mines to hinder attempts at clearance. Both UNITA and FAPLA used mines in this capacity.

B.   Ambush

Ambushes have been a common tactic throughout all phases of the Angolan conflict. The range of devices employed is dictated by the nature of the target, i.e. whether it is a vehicle convoy, an armored column or a foot patrol. Ambush has been the primary use of directional devices such as the M18A1 (Claymore) and MON-50 and MON-100. In most cases these devices were used in the remote-detonated role and therefore present no long-term hazard, as the remote detonators have been removed by the combatants. However, on other occasions, they have been deployed using tripwire initiation. There is evidence that remote mines have been left behind, with tripwires intact, after ambushes have been abandoned or have failed. These devices still present hazards (the case of one tripwire initiated mine that exploded killing a civilian and injuring another is given in the following chapter). Since many ambushes were set at key sites, such as wells, river fords, or track junctions, the incidence of remote devices at these locations is high. Few records have been kept of such minelaying.

C. Bridgehead Mining

All combatant groups in Angola have commonly employed the practice of mining bridgeheads. This now presents not only a danger to civilians, but a serious obstacle to economic and social rehabilitation. Bridgehead mining has taken two forms. One is the defensive mining of standing bridges. The second is mining to prevent the repair or reconstruction of destroyed bridges, by mining the damaged bridges, approach routes and adjacent river banks. Bridgehead mining is a particularly serious problem in southern Angola and in Moxico Province.

D. Defensive Mining of Key Structures and Facilities

This was used primarily by FAPLA to deter insurgent action against key economic installations and strategic locations. This tactic is well-illustrated by the use of anti-personnel mines at the base of electricity pylons. Major roads, railroads, dams, oil installations, diamond mines and water pipelines were also protected in this manner. FAPLA also laid protective and nuisance minefields around the perimeters of towns and municipal centers where UNITA attacks were expected.

In addition, FAPLA laid large defensive minefields around the towns of Cuito Cuanavale and Cassinga to counter UNITA advances in these areas. This specific use of minefields for defense against conventional ground assault was in accordance with Soviet military doctrine. At least some of these minefields appear to have been recorded.

E. Random Dissemination

The random dissemination of land mines can have an immediate military purpose, i.e. to deter infantry attack and reconnaissance patrols. It can also merely have the purpose of terrorizing civilian communities. However, after so many years of war, it is now impossible to determine whether the randomly disseminated mines in a particular area were originally deployed for one purpose or the other. In the post-combat era, any such distinction is no longer relevant. No party kept records of where such land mines were disseminated, and the mines were laid irresponsibly without regard for the welfare of the civilian population.

Use of Improved Explosive Devices and Booby-Traps

Improved explosive devices (IEDs) and booby traps were used by combatant forces. UNITA and South African forces are known to have made regular use of anti-lift devices on mines, which explode when an attempt is made to clear the mine. These devices are typically improvised from plastic explosive. An indication of the emphasis placed on improvised devices by South African units operating in Angola is illustrated by the fact that Zulu Force groups during Operation Savannah in 1975 carried as much as two tons of plastic explosive.[1]

Reports of booby-traps are frequent but mostly unsubstantiated in detail. However, Africa Watch was shown evidence indicating that improvised booby-traps have been employed. The evidence consisted of photographs of several "mousetrap" pressure-release switches and some tripwire devices linked either to plastic explosives, grenades or mines. Major M. G. Cox of the British army completed a survey of land mines in August 1991, which formed the basis of British army involvement in mine eradication in Angola. Major Cox wrote:

> The greatest problem facing UNITA mine clearance teams is that they booby-trapped a lot of their anti-tank mines. The traps are designed to defeat clearance operations. There are no records of their locations nor are there techniques for their safe neutralization. As little is known about these devices, Maj. Cox has secured a promise that inert working examples will be sent to the UK for analysis.

Cox went on to explain that he knew of two types of booby trap or anti-lifting mechanism, one based on magnetic influence, and the other on exposure to light.

---

[1] Breytenbach, *They Live by the Sword*.

However, a US army Major at the US Liaison Office in Luanda[2] told Africa Watch that he considered the problem of booby-traps had been exaggerated. He said, "those mines would have been laid a long time ago and I don't think anything too high-tech has been used here—it is all fairly basic technology, simple stuff." He did not deny that this "simple stuff" remained explosive. United States attitudes to the land mine problem have to be seen in the light of US denials that it provided land mines to UNITA (see below) and US non-involvement in mine clearance operations (see chapter five).

Most sources agree that high-technology anti-lift devices have been widely employed, especially by South African and UNITA forces. These include light-sensitive switches to prevent the removal of anti-tank mines (by detonating them when they are uncovered) and magnetic proximity devices (which are initiated by mine detectors). Again, the US army disagrees. The Major cited above said that the problem of magnetic proximity devices had been "exaggerated." He explained, "the magnetic proximity thing . . . came out of one incident which we have not been able to confirm. The mine could have detonated for any number of reasons—people hear the story and jump to conclusions." The Major maintained that battery-operated magnetic devices would have decayed over time and become inoperable.

Land Mine Records

Existing records on the locations of land mines are extremely scanty. Major Cox claimed that all FAPLA minefields were recorded and copies kept at local and national headquarters. He said that he had been given access to minefield records classified as "secret," which he considered to be "of the standard necessary to carry out a safe mines clearance operation." The document in question appears to be *Formulário para Campos de Minas*. However, such minefield maps only account for a small proportion of the total number of mines laid by FAPLA, most probably the defensive minefields laid around economic installations and

---

[2] Africa Watch interview, May 6, 1992. The meeting was scheduled to take place with Major Jay Garza, Jr., but his place was taken by another uniformed Major, who refused to identify himself but told Africa Watch that he was "as qualified to provide a briefing on mines as Major Garza."

important military bases. They do not include any of those randomly disseminated in the countryside.

The fact that FAPLA minefield maps remain secret may indicate contingency planning on the part of FAPLA military commanders in the event of a resurgence of hostilities.

The South African Defense Forces are known to have mapped some of the minefields they laid on their incursions into Angola, and may now be using these records in their clearance operations in the south of the country.

UNITA appears to have recorded very few of the mines that its forces laid. Major Cox wrote:

> In general, UNITA mines were laid randomly and without record. Their minefields were of the nuisance type designed to deny key routes and industrial mining facilities to the MPLA. Their most extensive mining operations were along major roads and all of the railways. In order to prevent easy clearance of these mines, UNITA extensively used anti-handling/booby trap devices.

The Cuban forces were also responsible for laying land mines. There are varying accounts of the practices used by the Cubans, some claiming that most Cuban minefields were accurately recorded, and others claiming that the Cubans kept no records at all.

Assessments of the Total Land Mine Threat

In the absence of comprehensive surveys, or even systematic sample surveys, no one knows the true extent of the land mine crisis in Angola. However, some estimates have been made of the number of land mines in Angolan soil, and the size of the operations needed to clear certain sectors.

Brigadier Fann Grobbelaar, a mines expert with the SADF, has said that some mines in Angola date back to the independence struggle against Portugal. "A mine in the ground is still lethal," he said. "The whole of Angola must be considered a mined area."

Colonel Bob Griffiths of the British Army Royal Engineers and chief of the British Military Mission in Angola, told Africa Watch that there "are twenty million mines in Angola spread over one-third of the land mass." He claimed that this figure was extrapolated from known supplies to Angola and other intelligence. Col. Griffiths further estimated that, of the twenty million mines, "four million are recorded and still in the ground, six million have been lifted or have functioned and eight million are unaccounted for." He could not give any details or explanation for his arithmetic, nor how these figures were arrived at, nor where the four million "recorded" mines were laid. He also estimated that 52,000 kilometers of roads had been mined, though he also claimed that most of these roads had been cleared—a highly contestable claim (see chapter five).

Griffiths went on to say that one hundred different mine types had been disseminated of which forty-six had been recorded. Africa Watch regards the presence of some of the forty-six as unproven. He was unable to explain on what basis the initial figure was estimated, but said that mines:

> were supplied by twelve different countries, of which Portugal, Russia, Cuba, South Africa and East Germany were the prime suppliers. There are also British and US mines and devices from many eastern bloc countries. Many mines are thought to have been supplied through individual commercial deals rather than as government to government support.

According to the Joint Mine Clearance Committee, between June and September 1991, a total of 22,124 mines were cleared.[3] The numbers have now risen. Major-General Helder Cruz, the senior Angolan army officer responsible for mine clearance, claims that 50,000 mines were cleared by joint FAPLA/FALA teams in the year after May 1991. He warns that "hundreds of thousands" of mines remain, which could take up to twenty years to clear.[4] Cruz doubted whether all the mines could be found and cleared but is reported to have said that "there is a good chance of destroying most of them before the September elections." This was an unduly optimistic claim. Cruz also said that the South Africans have supplied maps showing where the SADF laid mines.

---

[3] *Journal de Angola*, November 10, 1991.

[4] Reuters, June 12, 1992.

Of 330,000 mines believed to have been laid in the municipality of Cuito Cuanvale, Cuando Cubango province, the location of only 80,000 is known. The remainder of the mines have not been recorded or mapped.[5] The severe mine problem in this area forced the government and UNITA sappers to suspend mine clearing in early 1992, until new clearance techniques were developed and the sappers were retrained. Official figures up to mid-1992 indicate that at least eight soldiers were killed and twenty-five injured in clearance operations. The true figure for casualties is probably considerably higher.[6]

United States Involvement

The United States government and Congress have been significant though inconsistent supporters of UNITA, and have provided financial and military support. At least seven types of US-manufactured mines are present in Angolan soil. Major Cox of the British army noted that "the mines laid by UNITA forces were mainly from the USA." He did not, however, say who was the immediate supplier of mines to UNITA. His fellow British officer, Col. Griffiths also declined to characterize the US as a major direct supplier of mines.

At this writing, the United States government has not accepted that it bears any responsibility for the large number of US-manufactured mines in Angola.

---

[5] *Jornal de Angola*, July 28, 1991.

[6] David Sogge, *Sustainable Peace: Angola's Recovery*, Harare, 1992, p. 89.

3. THE HUMAN COST

Angola has one of the highest rates of land mine injuries per capita in the world. For a population of about nine million, it has tens of thousands of amputees, the great majority of them injured by land mines. The government claims that there are 55,000 amputees in the country. The International Committee of the Red Cross (ICRC) has a more conservative figure of 15,000, but that refers to lower-limb amputees only, excluding those who have lost an arm or their sight or who have been otherwise maimed or disfigured by land mines. Even taking the extremely conservative figure of 20,000 people seriously maimed by land mine injuries, this implies a rate of injury that would be equivalent to 500,000 people in a country the size of the United States.

The government has produced figures only for mine fatalities among FAPLA soldiers: between 1975 and 1991, 6,728 were killed. But in reality, there are no reliable estimates for the numbers of people killed by land mines. Because of the scarcity of medical care for the civilian population, the true total figure must be very high.

The material in this chapter is derived from several sources. One is interviews carried out by and on behalf of Africa Watch in two visits to Angola, in 1990 and 1992. Information about forty-seven land mine incidents was obtained from interviews on the first visit, including specific information on twenty-six victims. Forty-five victims were interviewed on the second mission, in Kuito-Bíe, Huambo, Viana in Luanda Province and Luanda city. It is also based on information provided by other sources, including the ICRC, which has an extensive program of assistance to land mine victims. A survey of 113 land mine victims from eight provinces carried out by the ICRC in 1990 provides some of the best information available.

The chapter first recounts some of the circumstances in which Angolans fell victim to land mines. It then goes on to recount the medical care and rehabilitation that is provided, and some of the problems affecting land mine victims in their attempts to live a semblance of a normal life.

Who are the Victims

The sample sizes of the three surveys are all small, and the sampling not statistically rigorous, but the cross-section of victims represented gives an indication of the Angolans who have suffered and continue to suffer from land mine injuries.

The majority of those interviewed originate from just two provinces, namely Bíe and Huambo. This is partly because the sample over-represents these areas, because of ease of access and the concentration of facilities for the disabled in these areas, but it is also probably true that these provinces have suffered a disproportionate share of land mine injuries. However, the under-representation of people from the south and east, particularly Moxico Province, is a serious drawback, as the land mine problem is also very severe in these areas.

Many of the victims are soldiers. About half of those admitted to the ICRC center for amputees at Bomba Alto, near Huambo, are soldiers, and half are civilians. Fourteen of those interviewed by Africa Watch in 1992 were soldiers, and they could give details of many of their friends and colleagues who had been wounded or killed. A.C., a soldier interviewed by Africa Watch in 1990, estimated that during his four years in FAPLA, twenty soldiers stepped on mines from his battalion of 500 men. A.C. himself trod on a mine in 1988.

The great majority of the soldiers are of course young men. This means that overall, a disproportionate number of those disabled by land mines in Angola are young men, a fact which has contributed to the militancy of many amputees in demand of their rights.

Among the civilians, there is a wide distribution of casualties. Men and women of all ages are affected.

The evidence available to us suggests that most land mine casualties are adult men, though this may not necessarily be a true reflection of reality (see below). Twenty-three out of Africa Watch's 1990 and 1992 civilian sample of sixty-one were adult men. Fifty-six of the ICRC's sample of 113 (almost half) were adult men.

There appear to be fewer adult women casualties. Africa Watch's two surveys interviewed sixteen adult women out of a total of sixty-one, and the ICRC included twenty-eight women among 113 (i.e. about one-quarter). In December 1990, there were thirty-five women among 120 waiting to be fitted with prostheses at Bomba Alto center for amputees.

There are several reasons for distrusting these figures. One is that the figures are for *survivors* only, and there is evidence to suggest that more women die from land mine injuries than men. Women tend to be more severely injured than men when they step on mines. The ICRC survey found that whereas only one-quarter of male amputees were amputated above the knee (forty tibial amputations and twelve femural), the same was true for half of female amputees (nineteen tibial and nineteen femural amputations). The damage done by a blast mine is related to the size, design and deployment of the mine, and to the weight of the person affected; men being heavier than women tend to suffer less severe injuries. Extrapolating from this, it would be expected that more women than men would be killed outright by land mines, or subsequently die from the effect of severe injuries. There is some anecdotal information to support this (see case 2 below).

In addition, it is possible that there is a bias in the provision of health care for the sexes. Men may be transferred to hospital quicker than women, and may be given better treatment. This would help more men survive. Having recovered, it is possible that men are given priority by the family and community in access to rehabilitation and prostheses. This would mean that male amputees are more visible than female, and more accessible to those doing surveys.

Children are an important minority of those affected by land mines. The 1990 and 1992 Africa Watch surveys included twenty children (aged under eighteen) including five under the age of ten. The ICRC included twenty-nine out of 113 (about one-quarter). In December 1990, fourteen children were among 120 patients at Bomba Alto.

One incident in which a child was killed occurred when a woman, aged forty-two, stepped on a land mine, losing her right leg. Her son, aged three, was killed in the same incident. A boy aged ten was killed when he played with a land mine that he found on a path. A girl of seven was injured when she trod on a mine trying to flee a UNITA attack on her village, and a girl of eight walked on a mine on her way to school.

Old people, being less mobile, are less prone to land mine injuries. Africa Watch's two surveys included only four people aged over fifty. The ICRC survey classified all those aged over forty as "old" and included eighteen people in this category.

Where the Mines are Planted

The great majority of mine victims interviewed had been injured by anti-personnel mines. Africa Watch's 1992 survey found only two anti-vehicle mine victims from a total of forty five. The ICRC survey found that ninety-six of 113 victims had been injured by anti-personnel mines (84 percent). However, anti-vehicle mines typically caused many more deaths at a time; for example one mine that had been set off by a truck killed five and injured ten passengers.

The ICRC survey distinguished only three categories of place where mines were laid: paths, roads and in villages or towns. It found 69 percent of victims injured on paths, 15 percent on roads and 16 percent in inhabited areas. The Africa Watch 1990 and 1992 surveys identified the locations of mines in more detail.

From a total of fifty-seven cases of civilian injuries, three occurred in fields. One example was F.G., a twenty-nine-year-old farmer from Cuito Cuanavale.

Case 1. F.G. stepped on a mine in October 1990. UNITA had often laid mines close to the river, about half a kilometer from his land, but no mines had ever been laid in the fields before. At 8:00 a.m. F.G. stepped on a mine in his own field. He thinks it had probably been laid by UNITA the night before. He was not alone in the fields, and his fellow workers carried him to the village in a cart. He stayed there for several days—he thinks a total of six—before being taken to hospital, first by cart and then by car. His left leg was amputated above the knee.

A total of thirty-five injuries happened on paths. People were walking to the field, school, market, or medical center. Often those injured were walking at the head of a line of people in single file, but sometimes they were in the middle of a line of people—those in front had safely passed over the mine, or the mine had a timed explosive device. Sometimes, villagers have suffered a spate of land mine casualties within a short space of time, usually after returning to the village following a UNITA attack, or after there had been a military presence in the area. This occurred in the following case.

Case 2: C.B., aged sixteen, was injured by a mine on a path to the fields in Caconda municipality, Huíla, in February 1989. On the same day that he was injured, six others were also hurt by mines. They were not together but on different paths in the same area. Of the six, three women died and three men were wounded. Soldiers passed on patrol on these paths, as well as civilians. They think that UNITA was the one that mined, in part because they saw UNITA-type boot marks around the area where the mines were.

This has become so familiar to villagers that fear of land mines has led to the wholesale desertion of villages.

Case 3: The villagers of Bave, Huambo Province, fled to Chipipa in August 1990 following a UNITA attack. FAPLA would not allow the villagers to return because FAPLA itself did not return. The villagers were finally allowed to return to look for food because they were growing hungry in Chipipa. Many went back on October 23, 1990, with a military escort. On the day of this return, however, a woman aged forty-five, a mother of seven, was wounded in the left leg when she stepped on a mine on a small path from the village to the fields. She was the only person on that path at the time. The villagers then all returned to Chipipa, because they were frightened by this injury, and because UNITA had burned all their houses.

Sometimes, land mine accidents happen wholly unexpectedly. This was the case for the injury to J.K., aged twenty-nine.

Case 4: J.K. was injured by a mine on October 25, 1990, in Chavola, near Calépi in Huíla Province. He stepped on the mine when he was hunting in the bush about four kilometers from his house. He had never heard of anyone stepping on a mine there or in his village. He did not know who had left the mine. His left leg had to be amputated at the knee.

While most mines on paths are buried, some are triggered by tripwires. A.Y., aged nineteen, described the explosion of a tripwire-initiated land mine, laid by unknown persons in Sanga village, close to the municipality of Mungo in Huambo Province. The mine detonated on October 18, 1989:

Case 5: A.Y. and his 17-year-old cousin, P.C., were walking on a path only used to walk to the fields. A.Y. was in front, when he kicked a wire that he did not see. There was an explosion, and his cousin took the full impact of the mine. He died immediately. A.Y. was injured in his right leg, which fortunately did not have to be amputated.

Various types of mine, especially scatterable mines, may be left on the surface of the ground. A boy of ten was killed by a mine left above ground in the village of Vionga Baixo, Bíe Province, in 1989. L.C., an older man from the village, described the context of the death:

Case 6: There were eight UNITA attacks [in 1989]. The pattern was that when UNITA attacked, FAPLA would withdraw. UNITA would sleep the night in the village and leave mines behind when they departed. When FAPLA reentered, soldiers would step on the mines. UNITA would place mines on the roads where the soldiers passed. The problem was that the people were also using the same roads: the larger roads were used for vehicles and the smaller roads were used for cattle.

There was one FAPLA officer who died stepping on a mine. Other FAPLA soldiers stepped on mines on a [non-asphalted] road to the village, but luckily no civilians stepped on mines there. But cattle were killed on mines. There were no mines in the fields.

A boy of ten died from a mine in 1989. A mine was left on top of the ground on the path where the boy was herding goats. He started playing with it and it exploded, killing him. We believe that it was a UNITA mine because UNITA had attacked two days earlier and had just pulled out, and FAPLA had not yet returned.

In its survey, the ICRC distinguishes commonly-used paths and infrequently used ones: 79.5 percent of accidents were on often used and 6.4 percent on rarely used. (The remainder were classified as unclear.) About two-thirds of accidents occur less than five kilometers from the village or town, giving the lie to the notion that there is a "safe zone" close to habitations.

Roads and roadsides are the second most common sites of land mine injuries. Both anti-vehicle and anti-personnel mines are responsible. Eleven out of the Africa Watch samples had been injured on the roads, nine of them by anti-personnel mines. Eleven of the seventeen people in the ICRC survey injured on the roads had been injured by anti-personnel mines. The greatest danger of anti-personnel mines is at the roadside, affecting people who left the road to follow a short cut, to rest or to urinate. Anti-personnel mines may also be laid close to anti-vehicle mines to deter clearance operations.

Case 7: A.J., a twenty-one year old farmer from Zaire Province, was on the way to market in 1990, when she left the road to enter the tall grass, and stepped on a mine. She screamed for help and people came to take her for some first aid. She was transferred to hospital the next day. At first the doctors tried to save her foot, but it began to putrefy and had to be amputated.

Case 8: C.C. stepped on a mine in Cangombe, near Bailundo in Huambo province in 1985, when she was aged nine. She was doing domestic service in the house of her aunt because the year before both her parents died while fleeing an attack; they were shot outside their house. C.C. was alone, walking from Longongu to Bailundo at about 2:00 p.m. when she stepped on the mine. She was on her way to the fields to look for thatch, walking along the side of the dirt road used by trucks. She does not know who laid the mine.

Many mines have been laid in built-up area. Africa Watch interviewed five people injured inside towns and villages. The ICRC survey included eighteen people injured in this way, consisting of:

- six on paths inside villages or towns;
- three in the gardens or yards of houses;
- three at entrances;
- three inside houses;
- two close to a well or water point;
- one on a football pitch.

Case 9: A.S., a forty-one year old man returned to his home at Camanongue, fifty-two kilometers from Luena, Moxico Province, three weeks after the ceasefire in June 1991.

> The day after returning home he stepped on a mine outside the door of a bar in the center of town. FAPLA teams had cleared the roadways of mines, but had failed to clear all the verges and areas adjacent to buildings.

Case 10: M.C., an eight-year-old girl, was walking to school inside the small town of Cuima, Huambo Province, when she stepped on a land mine. Others told her that it had been laid the day before by UNITA forces.

Case 11: M.D., aged fifty, was injured by stepping on a mine on October 8, 1989, in Chicuma, Benguela Province. His left leg was amputated below the knee. The mine was inside the village, near a house, on a path. He stepped on it at 11:00 a.m. on a Sunday morning as he was returning from church. This occurred about a week after the last UNITA attack, during which there was fighting that lasted many hours. Many died and many houses were destroyed. When FAPLA withdrew, UNITA entered and was there for four hours before leaving leisurely at 11:00 a.m., robbing the civilians' food.

Many mines have been planted on riverbanks, especially around bridges. As bridges and their approaches are a well-known location for mines, civilians tend to be very careful. The only person interviewed by Africa Watch who had been injured at a bridge had fallen off the bridge when it had collapsed, after which he trod on a mine.

Three others in the Africa Watch samples were injured at riverbanks.

Case 12: M.K., a twenty-four-year-old farmer from Cutato in Bíe Province, stepped on a mine in September 1988. While working in the fields, he walked to a nearby river for a drink. There had been no land mine injuries on the banks of this river so he believed that it was safe, and took no special precautions, but nonetheless he trod on a mine. Three days later, after his transfer to hospital in Kuito, his right leg was amputated. A FAPLA soldier later admitted FAPLA responsibility for the mine, saying that the area had been mined two years earlier. After this incident, the local people were warned. However, up until the time of M.K.'s injury, there had been no warnings, and no markings.

Africa Watch interviewed two people who were injured while running away from villages during UNITA attacks. One attributed the mine to UNITA, and one did not know who had planted the mine. However it is likely that many casualties that have occurred in this way have been caused by FAPLA defensively mining around population centers to deter UNITA attack. Though not set with injury to civilians in mind, these mines had precisely that effect when they impeded residents' flight rather than UNITA's attack.

Africa Watch interviewed three people who had been injured very close to FAPLA military posts (two on paths and one on a riverbank; all are included in the figures above). Though the victims tended to blame UNITA or say that they did not know who had planted the mines, it is likely that FAPLA units planted mines close to their posts in order to deter attack.

Some mines are left on or in the vicinity of the railroads. These are intended to disable trains, or to catch people who use the railroad and its embankments as footpaths.

There is a train for civilians running the short distance between Lobito and Benguela. It costs fifty kwanzas to ride and many civilians ride it. There are three empty cars in front, being pushed by the engine, in order to detonate any mines. Some passengers travel on these cars, as they are permitted to go free. There are about twenty-five cars behind the engine, loaded with people. The track has been mined, although it is not used by the military. People died in 1990 from the mining. In July 1990, two UNITA saboteurs were killed by their own mine as they were laying it on the tracks.

Who Laid the Mines

The majority of mines are attributed to UNITA. In Africa Watch's 1990 survey, twenty injuries and five deaths were attributed to UNITA, and twenty injuries and two deaths to "unknown persons." It may be that in some of those "unknown" cases, people knew who was responsible but declined to identify them. This would be especially the case if the perpetrator was FAPLA, as most interviews were conducted in

FAPLA-controlled territory. In many cases, however, the situation was genuinely too ambiguous to identify who was responsible.

In Africa Watch's 1992 survey, among a total of forty-five, six said that FAPLA was to blame (including one soldier blown up by a mine his colleagues had planted earlier), twenty-seven said UNITA, and twelve said that they did not know. Many of the "don't knows," particularly the six who were interviewed in Luanda, may have been reluctant to mention FAPLA.

The 1990 ICRC survey came up with a similar result. Eighty-three blamed UNITA (73.5 percent), fourteen blamed FAPLA (12.4 percent), one blamed the Cubans (0.7 percent), and fifteen said that they did not know (13.3 percent).

Knowledge about Minefields and Warnings Given

In very few cases were civilians warned that mines had been planted in a certain area. Residents relied solely on observing military activity and on the incidence of mine injuries to discover which areas were safe and which were not.

In Africa Watch's 1992 survey, twenty-three civilians spoke about their knowledge (or lack of it) concerning the localities of mines. Eighteen said they were not aware the area was mined, and no warnings had been given. After having been injured, many were able to work out when the mine had been laid and by whom, using circumstantial clues, such as the presence of certain forces in the area recently, boot marks close to the site of the mine, etc. Some of the mines had been laid the day before the injury, and others had been laid months or years earlier. During a reconnaissance patrol in July 1990, one twenty-four-year-old FAPLA soldier stepped on a mine that his colleagues estimated had been planted by UNITA in 1978.

Of the other five civilians, one was a newcomer to the area and so, he said, could not be expected to know which areas were mined, two were injured while using a path that they knew was often mined, one feared land mines on the path he was using but had no specific warning, and one man, who stood on a mine next to a bar, said that FAPLA forces had cleared the surrounding roadways of mines, but had not completely cleared the verges.

Africa Watch also interviewed thirteen military victims. These included Z.A., a FAPLA sapper.

Case 13: Z.A. lost a leg after stepping on a mine laid outside an electricity station at Dande dam, Bengo Province, in 1988. He was clearing the mines laid by his colleagues three years previously to protect the electricity station from attack by UNITA. Nonetheless, UNITA succeeded in destroying it in May 1988, and a FAPLA demining unit was sent to clear the earlier mines, presumably in order to facilitate repair work. The troops who originally laid the mines verbally warned the local population, but failed to provide any markings or maps.

Emergency Care for the Injured

For most of those injured by land mines, first aid was available within a few hours. According to twenty-two civilians questioned about this by Africa Watch, the average time that civilian victims waited for first aid was just under two hours. The maximum was six hours. For soldiers, assistance was usually more rapid, with immediate evacuation often by helicopter or vehicle. However, there was one case of a soldier who was given emergency help by his colleagues in the bush but then had to wait three days for evacuation. The ICRC also ran an air evacuation service from remote areas. Since the end of the war, air evacuation services have ceased, and those injured must rely on overland transport.

First aid for mine victims is usually extremely rudimentary, consisting of no more than bandaging the wound and providing comfort and perhaps some painkilling drugs. Transport to the nearest first aid post usually involved being carried manually or by cart; onward transport to hospital was usually by car or sometimes by airplane.

Civilians had to wait on average for about thirty-six hours before arriving at hospital. Three days was not an unusual wait, and one man believed that it had been six days before he received hospital treatment.

Medical Care and Rehabilitation

Care and rehabilitation for FAPLA soldiers has been the responsibility of the Serviço de Ajuda Médica-Militar (SAMM) of FAPLA. It functions very well, in part because the government and military are able to attract good people because they offer benefits and access to goods. Payment in money is little incentive in Angola because of the depths of the economic crisis.

Civilians receive treatment in civilian hospitals. Like everything else in Angola, adequate treatment is scarce. Drugs are often in short supply, and the staff are less well qualified and less well motivated. When the Cuban troops departed following the December 1988 agreement, the Cuban civilians left too. Among their number were many health personnel, who provided a highly professional service, and their departure left a significant gap in the medical services.

The variable quality of medical care has meant that hospitals can be dangerous for amputees. Wounds may become infected and there may be need for secondary or even tertiary amputations. There has also been a high incidence of osteomyelitis, a bone-wasting disease, which may set in after a poorly-done amputation.

The existing facilities for rehabilitating land mine victims are grossly inadequate. The ICRC has run its center at Bomba Alto, near Huambo, since 1980. This includes eleven technicians working solely on the manufacture of artificial limbs and seventy-eight workers in all.

Injured people come for a five-week period to Huambo and are lodged there at the Red Cross shelter. Ironically, those working and receiving limbs and therapy at the Huambo shelter cannot freely go to Bomba Alto, ten kilometers away, because of the threat of land mines on the road. At the center they are fitted with a prosthesis. They are measured and wait for it to be made, while receiving physical therapy. When the prosthesis is ready and adjustments are made, they practice with the new limb.

Angolan students are trained to become technicians at Bomba Alto; others study at the orthopaedic school in Huambo. Artificial feet and limbs are made from wood. Though there are woods nearby, it is not safe to enter the forests to cut the wood, because of land mines. Hence the wood used has been brought in from Cabinda.[1] Other raw materials such as resin and nails are hard to find because of the economic situation of the country.

Between January and November 1990, 631 new civilian and military patients were fitted with prostheses at the center. In total, 1,127 prostheses were manufactured in 1990, and 1,039 major repairs to prostheses were made during the same period. The ICRC also has a center at Kuito, in the UNITA-controlled areas of the southeast, which

---

[1] Due to the intensification of separatist activity by dissident factions of the Frente de Libertação do Enclave de Cabinda (FLEC) this source is no longer available.

produces nearly two-thirds as many prostheses. The ICRC would like to hand both centers over to the Angolans in due course. However, due to lack of money and organization in the government, there is little prospect of these centers reaching the stage where the ICRC can scale down its input or withdraw altogether.

The Swedish Red Cross runs an orthopaedic center at Neves Bendinha and the Dutch Red Cross has one at Viana, Luanda Province.

In its Jamba headquarters, UNITA's Special Department for War Wounded was set up in 1989. It has at least three units catering for war amputees. One of these is said to produce twenty artificial legs per month.[2]

A prosthesis can only be expected to last two to three years, and children require a new one at least every year, as they outgrow the old one. This means that a total of over 5,000 new prostheses is required every year, merely to cope with the existing number of amputees. This is more than twice the number currently being manufactured.

Social Rehabilitation

Angola remains a desperately poor country in which few facilities are available for the physically disabled. Most amputees are reluctant to leave the relative comfort of rehabilitation centers. Their future will consist of being cared for by their families, or attempting to earn a living in one of the few occupations open to them, such as street trading or—for those with education—secretarial work. The majority who come from farming backgrounds are likely to remain a burden on their families for the foreseeable future. Many have been reduced to begging; amputee beggars are already a common sight in Angolan towns.

The difficulties faced by female amputees are particularly severe. These are illustrated by a woman from Kwanza Sul, named Catarina:

Case 14: Catarina stepped on a land mine at the entrance to her field in February 1986. She was aged twenty-four, and pregnant at the time. One leg was severed in the explosion and the other injured. Although she cannot

---

[2] Steve Toussie, International Rescue Committee, "War and survival in Southern Angola: The UNITA assessment mission," typescript, no date, pp. 31 and 37-8.

work, she considers herself fortunate because her husband has continued to support her. She remarked:

> "Is it difficult [for a disabled woman] to find a husband? Yes! A man runs away when he sees that you don't have a leg. Or it has to be a man who also lacks an arm or a leg."

Maria, aged thirty-two from Bengo Province, was less lucky. She lost her leg in 1989 after stepping on a mine. She spoke to Africa Watch at a market place south of Luanda, where she was selling vegetables:

Case 15: "It is a difficult life, because I cannot do all I need to do. The family members help me, but my husband found himself another woman. I am here to sell, to say 'Amiga, amiga, buy this!' because it is necessary to earn something so that I and my children can live."

Life is difficult for amputated men too. J.D., a twenty-year-old former soldier, stepped on a mine shortly after entering military service. He was despairing about his future:

Case 16: "Working with a hoe isn't possible for me any more. It has to be work writing . . . I've passed fourth grade, and I would like to work at a company, non-manual work, sitting at a chair, writing. I don't have any hope for the future. I don't plan to marry; without a leg they won't accept me, as a woman wants a man who can work. Marry an amputated woman? That isn't possible either—how could we live?"

Amputees in Angola have become increasingly militant, seeking to draw attention to their plight and their perceived neglect by the government. Former soldiers have become especially vocal. The reasons for this were explained by Vincent Nicod, chief ICRC delegate in Luanda:[3]

---

[3] Interview with Africa Watch, May 7, 1992.

They feel that they have given a limb for the cause, for their country, and now there is very little interest in them. Conditions for them are difficult and when they leave the *abrigo* [shelter] there is very little for them. While they are there they are fed, they have shelter. There is a problem when the *abrigo* becomes full; many of the amputees there are equipped and should leave, but they stay and this means we are unable to take more patients because there is no more room. Demonstrations by amputees are becoming more common. During the war they tended to be kept in one place but now they are within the community. It is not surprising that they are becoming more militant.

Angola will have to live with the human cost of the land mines war for many years to come.

4. THE SOCIAL AND ECONOMIC IMPACT

Land mines have a significant impact on most areas of Angola's society and economy. There are tens of thousands of handicapped people. Thousands of acres of farmland, pasture and forest, and thousands of miles of riverbanks are unusable. For example, the fertile Mavinga valley in Cuando Cubango Province of southeast Angola is largely abandoned because of the large number of mines laid there by UNITA and SADF. Roads and paths cannot be travelled, rivers cannot be crossed, either by bridge or ford. The return of refugees is particularly hazardous. Commerce and movement are obstructed, and relief supplies can only be delivered with great difficulty. The eradication of land mines is an essential prerequisite for peace and economic development.

The nature of the war in Angola has made the social and economic impact of land mines particularly severe. For the most part, it was not a positional war, with fighting confined to specific heavily militarized areas. At one time or another, almost every part of the country was affected, as the foci of battle rapidly shifted backwards and forwards. Because of this mobility, the disruption of land communications was a major aim of UNITA, and the mining of roads, paths and bridges was consequently an important strategy. UNITA's strategy aimed at destabilizing the government by making any semblance of normal life impossible in as many parts of the country as it could. Outside its base area of the southeast, it consolidated and administered few areas. Instead, it sought to deny the government free use of these areas. The wide dissemination of land mines was a central part of this strategy. Meanwhile, FAPLA laid mines to try to prevent UNITA forces operating throughout the country.

Many of the land mines were therefore planted as part of a strategy with the deliberate aim of causing social and economic disruption. They will continue to have this effect long after the end of hostilities.

This chapter does not attempt to describe the complete social and economic impact of the land mines disaster in Angola. Instead, it concentrates on a single issue, namely the safe return of refugees from Zaire and Zambia. This illustrates in a dramatic way the huge problems created by the presence of land mines.

Repatriation

The great majority of Angolan refugees originate from the eastern provinces, particularly Moxico and Lunda Norte and Lunda Sul, and from the north, particularly Malanje and Zaire Provinces. In terms of density of land mines, these provinces are probably not worse affected than, for instance, Bíe and Huambo. However, the problems are quite severe enough to present major obstacles to the repatriation program envisaged by the UN High Commissioner for Refugees (UNHCR).

Virgilio Mora, UNHCR Repatriation and Logistics Coordinator, told Africa Watch:[1]

> The repatriation was scheduled to begin this year but no date has been set, due largely to the failure of Zaire, Zambia and Angola to sign a tripartite agreement [until March 1992] and also due to the rainy season. However, there has been some spontaneous movement and UNHCR has been assisting these people. There are essentially two "fronts"—the north, where the situation can be handled, and the east, primarily Moxico Province, where the UNHCR faces a nightmare scenario.

Mora did not believe that repatriation was possible in time for the September elections. He did not believe that enough was being done by the international community, concluding that "I fear, sadly, the Angolans must face many years of amputation."

Moxico and the Lundas

The eastern provinces of Moxico and the Lundas are very heavily infested with land mines, presenting almost insurmountable problems for the return of refugees. The problems exist on the roads, at bridges, and in the towns, villages and fields of the returnees. Some parts of Moxico are so heavily mined that not even UNITA or FAPLA allow their men to go there. These include Caripande and Kavungo regions of Alto Zambeze municipality.

---

[1] Africa Watch interview, May 4, 1992.

Of twelve key roads in Moxico, two are demined, two in process of demining (from Cazombo to Calunda and Luena to Kachipoque), and eight are out of use due to the presence of mines. Because of mined roads, there is no surface access to Lumbala N'guimbo or Cangamba. The latter has a population of about 65,000, which can only be served by air. Other concentrations of mines are at:

* Caripande to Cazombo to Caianda to Jimbe: vital roads for refugees coming from Zaire and Zambia, but heavily mined and currently out of use.

* Luau to Muconda road remains mined.

Leo Pavillard, senior UN coordinator for Moxico, Lunda Sul and Lunda Norte Provinces told Africa Watch:

[All roads] are assumed to be mined unless they have been cleared. . . . Before Angolan refugees return from Zaire and Zambia, all roads, river banks and bridges must be completely clear of mines. Otherwise the returnees will pay a high price in loss of life and severe injuries. It is not possible to quantify the risk.

The mining of the roads means that it is difficult and dangerous for refugees to return home by foot or vehicle, and when they have returned, it greatly restricts their economic opportunities.

In Lunda Sul Province, Muconda municipality, the situation is equally severe. This area illustrates the serious problems caused by the sabotage and mining of bridges. There are thirty major or strategic bridges down, and fifty-eight secondary bridges destroyed. Most of the damaged or destroyed bridges are surrounded by minefields, that must be cleared before the bridge can be repaired.

In 1992, surveys were made of bridges on the road from Saurimo to Muriege. The first major bridge over the river Mombo at thirty-seven kilometers from Saurimo had collapsed and was covered by about two-and-a-half meters of water. The government civil engineer accompanying the survey said that a temporary wooden bridge could be built about sixty meters downstream; the UNITA engineer said that the area would have to be demined before any work could begin.

In another incident, eighty-four kilometers along the Saurimo-Cacolo road, at the river Luangue, work started on the construction of a steel bridge but a worker was injured by an anti-personnel mine. Work then ceased until the area around the site had been demined.

In general, work on rebuilding the bridges can only begin when the roads have been cleared of mines. This means that fords have to be used to cross the rivers. However, many river banks are hazardous due to mines. The mining of bridges is therefore in every way as serious a problem as the mining of roads, and equally an obstacle to the return and reintegration of refugees.

In March 1992, the government claimed that in the provinces of Moxico, Lunda Norte and Lunda Sul, 4,000 had been deactivated, together with 377 "heavy armed projectiles," 278 hand grenades and 27 bombs.[2]

The demining effort has to date concentrated on roads, bridges and built-up areas. Clearing footpaths, fields and riverbanks is a lower priority. However, if the refugees are not only to return, but to reestablish their former lives, it will be essential for land to be returned to the community by means of eradicating these land mines.

Zaire, Uíge and Malanje

The majority of refugees in the north of Angola originate from Zaire, Uíge and Malanje provinces. In this area, the problems posed by land mines are markedly less severe than in the southeast, and the repatriation of refugees has proceeded more smoothly. Minesweeping operations ended along the major roads in the provinces in early August 1991. In the period June-July, twenty-five anti-personnel mines and two anti-tank mines were deactivated.[3] However, the presence of land mines still poses a significant threat to any prospective returnees.

Some of the larger remaining land mine concentrations in Zaire province include:

* M'banza Kongo- Luvo border post mined with anti-personnel mines.

---

[2] *Jornal de Angola*, March 26, 1992.

[3] *Jornal de Angola*, August 14, 1991.

* Nóqui-border area mined with anti-personnel mines.

* Tomboco to Lufico road mined.

* Maquela do Zombo, Huito to Banza Sossu mined with anti-personnel mines.

* Béu-Fiscak border post area mined with anti-personnel mines.

* Quimbele and surrounding area mined with anti-personnel mines.

* Mococola-Santa Cruz-Massau mined with anti-tank mines.

* Quifuata-Cambamba road mined.

The extent of the mining of roads has meant that much of the repatriation effort has had to be carried out by air, which is expensive.

In Malanje Province, there are many concentrations of land mines, which continued to deter people from returning to their home areas, even before the outbreak of hostilities in November 1992. In October, shortly after the elections, Julia Kissanga, a peasant woman, described conditions in her municipality:

> Here in Massango we are surrounded by mines: all the fields around our town are mined. A friend of mine died by a mine a while ago. I spend much more time getting water from the river, as the old path is too dangerous. Some mines around our washing place were cleared by the soldiers—they wanted to wash too, but they do not have to carry the water! My family has fled to N'dalatando, Malanje and Luanda. These mines make this no safe place for children. They cannot play safely and our fields are far away. They are safer in the cities.

Other concentrations of land mines in Massango municipality are in Dala Samba and Quingeungue. Elsewhere in Malanje Province, mines continue to pose problems at:

* Marimba; Quela; along power lines Lombe-Kalandula; along the verge of the Cacuso-Capanda road.

* Strategic installations, including the airport. When a South African clearance team set up camp in July 1992 at the airport they swept a public area previously assumed to be safe and found two anti-personnel mines.

* In Malanje's southern municipalities there are large concentrations of land mines around Capunda, Quimbango and Quirima. The main roads in the province were cleared in June and July 1991 by a joint FAPLA/UNITA brigade. A total of 320 mines were deactivated.[4]

In a peaceful Angola, Malanje Province would be a center for tourism. However, even at the famous Kalandula waterfalls, mines are a problem. A former resident of Kalandula, who used to work in the hotel overlooking the falls, told Africa Watch in September 1992:

> The access to the hotel is mined, which is why we have made no attempt to return and rebuild it. It is a shame, we could be making lots of money from the UN and other foreigners; instead we sit in Malanje city drinking, dreaming and talking about better times.

Game parks also suffer from mines. In the Bicuar National Park in Huíla Province, wandering elephants have set off anti-tank mines.[5] Even after the end of conflict in Angola, such dangers put into question a rapid restoration of game park tourism.

---

[4] *Jornal de Angola*, August 1, 1991.

[5] *Jornal de Angola*, December 27, 1991.

Overall Impact

Geoffery Winfin, the UN Emergencies Coordinator in Luanda, counts land mines as the UN's "single biggest problem."[6] In an interview with Africa Watch, he explained in more detail the nature of the problem, and why there was insufficient attention to it:[7]

> I would say that demining is an integral part of the peace process and therefore it should be included, as far as funding is concerned, into that process.
>
> My feeling is that to some extent the demining has been considered as a marginal part of the peace process, which, in my opinion, is not a reasoned approach. If you think in terms of the electoral process, if you think in terms of the movement of people, the problems of refugees and *deslocados* [displaced people], you realize that the mines problem is at the center of all those issues. Therefore I am a bit surprised that more attention has not been given to that problem.
>
> I realize that recently some countries have agreed to give assistance on a bilateral basis for mine clearance. But if you compare the development of the electoral process, you must come to the conclusion that the demining part of that process is lagging behind, when, if you look at things logically, it should lead the process in order to facilitate the other actions.
>
> I think it is interesting to examine the different situation that exists in Angola in comparison with Cambodia. In Cambodia the United Nations is responsible for the implementation of an electoral process. It could be said that it is a less difficult situation for the United Nations than exists here in Angola, where we are not responsible

---

[6] "Angola—A Very Fragile Peace," *Time*, January 6, 1992.

[7] Africa Watch interview, May 7, 1992.

for that process. Our role is, first, to give technical assistance to the electoral process through UNDP, secondly to observe the process and verify that the elections will be conducted in a fair and just manner. So we start with the basic principle that the Angolan government is responsible for the electoral process rather than the United Nations, we take a prudent role, supporting the process, but we are not here with a mandate to implement that process in place of the proper authorities. Therefore when we look at some of the delays that are becoming a problem, which give us great concern, what we can do as an observation organ is to express our concern to the Angolan government that the process is not being implemented to the United Nations' satisfaction.

Which brings us back to demining. In Cambodia it would be for the UN to make sure that the job gets done because they are responsible for the whole process. That is not the case here in Angola. We can only call the attention of the government to the slowness of the demining operations. In fact the government have been negotiating with the bilateral donors to obtain assistance with the demining.

5. MINE CLEARANCE INITIATIVES

There is a remarkable contrast between the widespread recognition that land mines present an extremely serious threat to Angola, and the actual response to the challenge of eradicating the mines during the pre-election period, when relative peace prevailed and clearance initiatives were possible. There has been no systematic assessment of the extent of the land mines problem, nor any serious attempt to coordinate eradication in a systematic manner. The clearance attempts by FAPLA and FALA teams were inadequately supported in both technical, financial and logistic terms. In addition, there is at least one example of an initiative cynically based on potential profit and future business opportunities, predominantly arms deals.

There have been several separate initiatives to clear land mines currently in Angola.

FAPLA/FALA Teams

FAPLA/FALA teams consisted of soldiers from both armies. During the pre-election period, they were working throughout the country with varying success. Probably, their best chances of success lay in the south of Angola, where units from SADF were providing training and technical support to clear mines from the areas in which SADF was formerly active (see below).

FAPLA/FALA teams were using manual clearance methods, partly because of the lack of heavy equipment, and partly because they considered it the most effective. The priorities have been to demine the major roads and railways, and the interiors of towns and villages. However, questions have been asked about how systematically the verges of major roads have been cleared.

Leo Pavillard, UN senior coordinator for Moxico, Lunda Sul and Lunda Norte Provinces, told Africa Watch:

> [All roads] are assumed to be mined unless they have been cleared. Even those that have been cleared continue to be a risk because the mine clearing techniques are insufficient. There have been several cases of deaths on the roads that have supposedly been cleared.

Agricultural land and pasture did not appear to be a priority for the demining teams, except in the case of commercial plantations, where the owners are able to pay for private demining operations.

Overall, the national mine clearance program suffered from many problems. A British military assessment of mine clearing operations states:

> Mine clearing operations lack central direction and adequate resourcing. The Joint Mine Clearing Committee of the JVMC remains ineffective; its regional subordinate groups lack resources; there is no comprehensive strategy or program. As things stand now, the Angolans are not capable of gaining control of the situation and making any significant inroads into the problem. Any assistance to the Angolans is likely to be squandered.

The document goes on to state:

> In mine clearing as with most other military undertakings we have observed in Angola, a lot of talking is being done but very little else . . . There is no quick fix to this problem short of taking over the project from them.

Ulterior motives on the part of the British military may be detected in the concluding assessment. The capacity, competence and motivation of the British involvement will be examined below.

There is, however, a more general acceptance of the limited impact of the Angolan de-mining effort. This is largely attributed to lack of organization and support. At a meeting on March 4, 1992, the Joint Mine Clearing Commission identified the following problems:

a) Serious command, control and communication problems at all levels.

b) No mine clearing entity, national, regional or unit-level, has a radio.

c) None of the mine clearing teams nor regional mine clearing commissions has a vehicle.

d) There are no evacuation capabilities or arrangements for personnel injured during mine clearance operations.

e) Mine clearance teams lack equipment such as helmets, flak vests, mine markers, engineer tape and demolition materials.

f) There is insufficient detection equipment.

g) There are no batteries for detection equipment.

These problems persisted despite the involvement of British military teams in assisting FAPLA/FALA efforts.

SADF Involvement

The South African Defense Forces were active in providing technical assistance and training to FAPLA/FALA clearance teams in the south of the country. In mid-1992, there was a general acceptance from almost all sources that the South African contribution was a well-motivated project based on a good knowledge of the general problems and the specific devices (many of which were laid by the SADF itself). All the Angolan parties responded positively to the South African initiative. A military representative of the US Liaison Office in Luanda was more skeptical, however. He told Africa Watch that there was insufficient coordination and planning:

> I believe this was all done at a very high level, at the Foreign Minister level I think. I believe they have some of their training team here already but are having trouble finding Angolans to train. This was done unilaterally, it's not part of any overall plan—I do not think that is a positive step.

However, this was the only significant criticism of the South African effort that Africa Watch was able to obtain.

British Army Initiatives

Colonel Bob Griffiths, head of the British Military Mission to Angola, told Africa Watch that British government policy is to provide a neutral source of advice and assistance to the Angolan authorities. He said, "We have no interests here other than to see a stable situation in southern Africa."

British support to the Angolan mine-clearing operation has consisted of two two-man Royal Engineer teams dispatched to instruct members of the envisaged integrated New Angolan Army (FAA) in British minefield clearance techniques. The first mission was in December 1991, the second in March 1992. Some indications of the priority given to this training can be seen by reference to the post-deployment report of the second team:

a) The team undertook six-and-a-half hours pre-deployment training.

b) The original deployment dates were February 19 to April 4, but due to problems with the shipping of mine clearance equipment the team was twice delayed, only departing for Angola on March 19. The completion date, however, remained the same.

c) "The period between 20 and 30 Mar 92 was spent familiarizing the team with the local area and unloading/checking the equipment sent by the FCO [Foreign and Commonwealth Office]."

d) The students were assembled on March 30 ready to begin instruction the following day.

e) Training began at 10:30 a.m. on March 31. Eighteen students from FAPLA and FALA attended, of whom twelve had attended the previous course in December 1991. At 4:00 p.m. a further five untrained students from FALA arrived. The total training period was two-and-a-half days.

f) The course finished with a mine clearance "exercise." A "packed-earth sports pitch was available for occasional use. This proved ideal for the final phase of the trg [training] where four small minefields were set up to demonstrate their new skills to the press, TV and VIP visitors." This was followed by a presentation of mine clearance equipment to the Angolans.

g) At a forum that followed, Major Rock, a senior FAPLA representative, "regretted that the team were not able to see some of the problems first hand."

h) The team left, as scheduled, on April 4.

Most specialists would question whether training of this nature had any practical value at all, especially, as Major Rock so politely intimated, the trainees had a more intimate knowledge of live materials than their supposed trainers.

A restricted cable from the United States Liaison Office (USLO)[1] to the US State Department shown to Africa Watch was more outspoken regarding the British training program. It said that although the British had obtained good public relations from their exercise, "effectively all they had done was give the Angolans thirty-six sets of Austrian detecting equipment."

The extent of British expertise in mine clearing in Angola can be further questioned. The British military team's assessment of the nature of the problem and the criteria for successful mine clearance were explained by the team's head, Col. Griffiths. He told Africa Watch that 52,000 kilometers of roads had been mined in Angola. Of this, he said:

> Eighty percent has been cleared to what, in British military terms, we would classify as a forty percent proof factor. In other words, they are sixty percent short of total clearance.

Griffiths gave no details of his sources of information, nor his justification for such estimates. Other sources have questioned how some forty thousand kilometers of roads could have been cleared, given the limited extent of mine clearance operations and the lack of an organized plan for clearance. In addition, one specialist commented:

---

[1] Refusing to recognize the MPLA government of Angola, the US had no Embassy in Luanda.

A forty percent clearance factor is nonsense. If a road is cleared, there may be a five or possibly a ten percent factor of uncertainty—but sixty percent, that simply means that the road is not cleared.

A secondary aspect to the British involvement also deserves mention. The British policy was described by Col. Griffiths and his colleagues as being solely providing "neutral advice" to the Angolans. This claim does not stand up to close scrutiny.

According to Col. Griffiths, his "neutral advice" has so far included ensuring that the South Africans do not gain any commercial advantage from their assistance to the mine clearance teams in the south. He told Africa Watch, "we would cry foul if the South Africans try to charge Angola for their work." After a brief assessment of other initiatives, he said that "the whole clearance initiative will be under the control of a joint Angolan-UK national coordination body which will be funded by donors such as the EEC." Griffiths then went on to explain that the "major work" would be undertaken by a UK company and the South Africans. He said that he could not reveal the name of the UK company because it was commercial-in-confidence. Further sub-contracts would be awarded to independent companies.

His colleague later confirmed to Africa Watch that the UK company involved was Royal Ordnance, and Africa Watch was later introduced to a Royal Ordnance representative who said he could be contacted through the British Embassy. Royal Ordnance is the recently privatized British arms manufacturer that is the major supplier of the British army, as well as an aggressive promoter of arms exports. Col. Griffiths seemed unaware that anyone might question the "neutrality" of British advice when a company so closely linked to both the British government and the international arms market was involved.

United States Involvement

When Africa Watch requested a clarification of planned US involvement in mine eradication, the response indicated that the US does not plan to assist with mine clearance. A US army Major at the USLO in Luanda stated the US position clearly:[2]

---

[2] Africa Watch interview at USLO, May 6, 1992.

> The State Department definitely does not have any intention of any involvement in [mine eradication]. . . . I think . . . that this is too low down the chain, it is too small a problem for us to get an involvement—that is not to say that mines are a small problem for Angola, it's just not on the scale that we are likely to get into. Of course, that is unless we come under pressure internationally or at home, then things may change. I think that our Department of Defense does not have the same close links to commercial interests as, say, the British Ministry of Defence—our Administration tends not to act as a front, an advance contact for commercial interests in these matters.

To date there has been no initiative by the US government to train or fund demining activities in Angola.

Equator Bank, USA, Initiative

The Equator Bank has attempted to interest the Angolans in the S-TRON Cast System, an experimental ground comparison survey method. This would bear an approximate lease cost of $5.6 million for a minimum of ten sets for three months. Both the USLO and the British oppose the introduction of this system, citing its unproven status. Both offices also point out that the Equator Bank would be "using" Angola to test the equipment, at Angolan expense. A British document, assessing the offer, states:

> Rather than expecting the Angolans to foot such a steep bill for an experimental system, it may be better to bring just one of the sets here, to prove itself, and then ask the Angolans to find a way to pay for more sets. . . . The Benguela Railroad would be a suitable location to conduct such a test. According to Major Kapapelo, the railroad is functional from the coast to the town of Huambo, yet only the tracks themselves have been cleared. The accompanying right of way, which would normally be travelled by pedestrians from the villages that line the railroad, has not been cleared of mines. Not only would the task of detecting mines along a

railroad pose a challenge to any system, but, if successfully executed, it would also immediately benefit the Angolans.

While the proposal that the S-STRON Cast System should be required to prove itself in the field is a good one, the specific proposal is suspect. From a technical standpoint, it proves little to test such a system over an area where such specific problems exist as on the Benguela Railroad. The point of a survey system, if it is to be effective, is to identify mined locations over a wide area of dissimilar ground situations. Moreover, the high ore content in the ballast makes the use of mine detectors unreliable.

The S-STRON system is expensive. $5.6 million is a large amount to spend on a three-month survey. This money could fund the training, equipment and deployment of many survey teams for a year, or the provision of three thousand mine detectors, or the mounting of many training courses for Angolan clearance teams. It is indicative that the USLO also opposed the Equator Bank proposal. USLO told Africa Watch:

> The Equator Bank was acting on behalf of a company called S-STRON. The State Department opposed the project because the system isn't proven and we believe that they should bring it here and prove that it works before they get money from the Angolans for its use... . In any case, it is far too high-tech for this country, totally inappropriate. What is needed here is basic techniques supported by expertise.

The Cap Anamur Initiative

The German-based humanitarian organization Cap Anamur has been bringing Soviet-built T-55 tanks, decommissioned from the former East German army to Angola. The plan is to use them for mine clearance on roads in the south of the country. The tanks are to be fitted with an array of flails, rollers and ploughs to destroy or detonate mines. The USLO told Africa Watch that it "imagined that it is the first step in a plan for some future commercial involvement," though this is mere speculation given Cap Anamur's humanitarian credentials.

No details were made available to Africa Watch. No explanation was given why it made sense to ship unwanted European tanks to a country that already had a surfeit of tanks (including about 200 T-55s). Surely it could be argued that only the flails and other special equipment needed to be shipped.

However, the systematic clearance of the roads in south Angola is an important project. If it is supported by adequate logistics and expertise, the Cap Anamur project could prove to be a major contribution to land mine eradication. Some of the necessary support programs for the planned route clearance that are needed include the parallel survey of land made accessible by the clearance of the roads, and the marking or eradication of the minefields discovered. If this is not done, the road clearance program will not succeed in returning land to the community, and may even increase land mine casualties as civilians return on cleared roads to uncleared areas.

Conclusion

Major Cox of the British army concluded his study of land mines in Angola:

> Angola has great agricultural and natural resources potential. However, after sixteen years of civil war, much of the land, resources and arterial routes are cut off by mines. As people are still dying of starvation, the acceptance of casualties in mine clearance operations will continue.

This is a sober conclusion. The imperatives of returning refugees, delivering emergency humanitarian aid, and reestablishing the economic and political viability of Angola cannot wait. If mine clearing efforts continue to be conducted in the present inadequate manner, then not only will there be needless casualties during the clearance operations, but there will be continuing large scale casualties among the civilian population.

6. LAND MINES IN INTERNATIONAL LAW

Land mines,[1] unlike chemical and biological weapons, have never been banned. On the contrary, international law specifically permits the use of land mines to achieve military objectives. However, the 1981 Protocol on Prohibitions or Restrictions on the Use of Mines, Booby Traps, and Other Devices, otherwise known as the Land Mines Protocol, does contain restrictions on mine warfare which are designed to protect civilians.[2]

The Land Mines Protocol is not directly applicable to the Angolan conflict. It applies only to international armed conflicts and to some self-determination wars. Fewer than forty countries have ratified, accepted, approved or acceded to the UN Convention. The government of Angola has neither signed nor ratified the Land Mines Protocol. In theory, this means that UNITA is not bound by the protocol either. The status of the other main parties to the conflict and arms suppliers to Angola is as follows:

* Cuba ratified the Land Mines Protocol on March 2, 1987, together with the other two 1981 protocols.

* South Africa has neither signed nor ratified the Protocol.

* The USSR ratified the Protocol on June 10, 1982, together with the other two 1981 protocols.

* The United States has signed the Protocol (in 1982) but has not yet ratified it.

However, many of the provisions of the 1981 Land Mines Protocol are already a part of customary international humanitarian law

---

[1] Mines are as defined as "any munition placed under, on or near the ground or other surface area and designed to be detonated or exploded by the presence, proximity or contact of a person or vehicle. . . ." Land Mines Protocol, Article 2(1).

[2] This protocol, known as Protocol II, is one of three protocols annexed to the 1981 United Nations Convention on Prohibitions or Restrictions on the Use of Certain Conventional Weapons Which May Be Deemed To Be Excessively Injurious or To Have Indiscriminate Effects, UN Doc.A/Conf.95/15 (1980) ("UN Convention").

and thus binding on the parties to the Angolan conflict. The two key provisions in this regard are prohibition on indiscriminate use of mines and the obligation to minimize or avoid civilian casualties. In addition, it provides the basic framework whereby the use of land mines can be assessed.

The Basic Rule: Protecting Civilians and Civilian Objects

Under customary law, civilians and civilian objects may not be attacked. U.N. General Assembly Resolution 2444, *Respect for Human Rights in Armed Conflict*,[3] adopted by unanimous vote on December 18, 1969, recognizes several principles of customary law protecting civilians. It states in part:

> (a) that the right of the parties to a conflict to adopt means of injuring the enemy is not unlimited;
>
> (b) that it is prohibited to launch attacks against the civilian population as such;
>
> (c) that a distinction must be made at all times between persons taking part in the hostilities and members of the civilian population to the effect that the latter be spared as much as possible . . . .

The Land Mines Protocol was adopted largely in response to the large number of civilian casualties caused by mines and unexploded munitions in Vietnam. It derives its provisions from customary law principles, and among other things, requires that combatants take "feasible precautions" (defined as "practicable or practically possible") under the circumstances to protect civilians from the effects of mines and booby traps.[4] The parties are required to keep records of minefields so

---

[3] See second and third paragraphs in the preamble of G.A.Res. 2444, 23 U.N. GAOR Supp. (Wo. 18) p. 164, U.N. Doc. A/7433 (1968).

[4] Land Mines Protocol, Article 3(4).

that they can be cleared once hostilities have ended.[5] It prohibits in all circumstances the use of mines "either in offence, defence or by way of reprisals, against the civilian population as such or against individual civilians."[6] It also prohibits the use of land mines "in any city, town, village or other area" where civilians are concentrated, unless combat between ground forces is taking place or imminent in the area and the mines are placed around a military objective,[7] or measures such as putting up warnings are taken to protect civilians from the effects.[8]

Article 3(3) of the Protocol prohibits the indiscriminate use of land mines. It defines indiscriminate use as any placement of mines:

(a) which is not on, or directed at, a military objective; or

(b) which employs a method or means of delivery which cannot be directed at a specific military objective; or

(c) which may be expected to cause incidental loss of civilian life, injury to civilians, damage to civilian objects, or a combination thereof, which would be excessive in relation to the concrete and direct military advantage anticipated.

If a weapon cannot with any reasonable assurance be directed at a military objective, it is considered "blind" and, under Article 3(3), indiscriminate. Contact land mines are blind when left in an area through which civilians pass, since they can be detonated by civilians as well as fighters. Experts on the laws of war also state that "land mines, laid without customary precautions, and which are unrecorded,

---

[5] Land Mines Protocol, Article 7(1).

[6] Land Mines Protocol, Article 3(2).

[7] A military objective is defined as "any object which by its nature, location, purpose or use makes an effective contribution to military action and whose total or partial destruction, capture or neutralization in the circumstances ruling at the time, offers a definite military advantage."Article 2(4), Protocol on Prohibitions or Restrictions on the Use of Mines, Booby Traps and Other Devices (Protocol II).

[8] Land Mines Protocol, Article 4(2).

unmarked, or which are not designated to destroy themselves within a reasonable time, may also be blind weapons in relation to time."[9]

Therefore a mine not programmed to self-destruct and not removed from an area after fighting there has ceased becomes blind and therefore indiscriminate as well.

Prohibition of Disproportionate Attacks

The legitimacy of a military target does not provide unlimited license to attack it. The customary law principles of military necessity and humanity require that the attacking party always seek to avoid or minimize civilian casualties.

The Land Mines Protocol codifies the "rule of proportionality" in customary law as it relates to collateral civilian casualties and damage to civilian objects. It thus prohibits as indiscriminate any placement of mines "which may be expected to cause incidental loss of civilian life, injury to civilians, damage to civilian objects, or a combination thereof, which would be excessive in relation to the concrete and direct military advantage anticipated."

This two-pronged test of proportionality requires an assessment of the "concrete and definite military advantage" expected: such an advantage should be "substantial and relatively close," and "advantages which are hardly perceptible and those which would only appear in the long term should be disregarded."[10] Under this test, the possibility that enemy troops may at some undefined time in the future move across a certain path may be too remote and insubstantial to qualify as a "concrete and definite military advantage."

The second prong of this test is that the foreseeable injury to civilians not be "excessive" in relation to the expected military advantage. Excessive damage is a relational concept which requires a good-faith balancing of disparate probabilities, but there is never a justification for

---

[9] M. Bothe, K. Partsch, and W. Solf, *New Rules for Victims of Armed Conflicts: Commentary on the Two 1977 Protocols Additional to the Geneva Conventions of 1949 (Geneva: 1982) ("New Rules")*, p. 305.

[10] International Committee of the Red Cross, *Commentary on the Additional Protocols of 8 June 1977 to the Geneva Conventions of 12 August 1949 (Geneva: 1987) ("ICRC Commentary")*, p. 684.

excessive civilian casualties.[11]

These two factors must be weighed in good faith by the commanders responsible for mining. In our opinion, the possible military advantage of injuring a soldier or deterring movement of soldiers by a mine laid on a footpath, without warnings and left for an indefinite time, is not sufficiently "concrete and direct" to outweigh the likely injury to civilians. The objective of injuring a combatant will not be achieved if a civilian steps on the mine first.

Under this customary law rule as well, the practice of leaving unmarked, unrecorded land mines that do not self-destruct in civilian-traveled areas is indiscriminate.

Prohibition Against Starvation of the Civilian Population

By prohibiting starvation of the civilian population as a method of warfare or combat, Article 54 of Protocol I and Article 14 of Protocol II of 1977 to the Geneva Conventions of 1949 establish a substantially new rule which has been accepted by many governments as customary law,[12] and which imposes important restrictions on the use of land mines, especially in farming areas.

Article 14, Protocol II provides:

> Starvation of civilians as a method of combat is prohibited. It is prohibited to attack, destroy, remove or render useless, for that purpose, objects indispensable to the survival of the civilian population, such as foodstuffs, agricultural areas for the production of foodstuffs, crops, livestock, drinking water installations and supplies and irrigation works.

This prohibits starvation as a method of combat, "i.e., when it is used as

---

[11] *ICRC Commentary*, pp. 625-26.

[12] See Charles A. Allen, "Civilian Starvation and Relief During Armed Conflict: The Modern Humanitarian Law," 19 *Georgia Journal of International Law and Comparative Law* 1 (1989).

a weapon to destroy the civilian population."[13]

While recognizing that it is still permissible to starve the enemy army, the article imposes sharp limits on that practice. That "objects indispensable for the survival of the civilian population" may also be of benefit to the enemy army does not give license to attack them. The narrow exception is where the objects are specifically intended as provisions for combatants,[14] which is generally taken to mean foodstuffs actually in the hands of the enemy armed forces. It is not permitted to destroy or render useless agricultural areas for the production of foodstuffs because as a practical matter it is impossible to distinguish between the part intended for military and that intended for civilian use.[15]

Crops and agricultural fields may be attacked, however, when used in direct support of military action; the *ICRC Commentary* provides an example:

> What is the position if such objects hinder the enemy in observation or attack? This might be the case if crops were very tall and were suitable for concealment in a combat zone. . . . [I]f the objects are used for military purposes by the adversary, they may become a military objective and it cannot be ruled out that they may have to be destroyed in exceptional cases, although always provided that such action does not risk reducing the civilian population to a state of starvation.[16]

Neither party may destroy objects indispensable to the survival of civilians because it suspects those civilians of supporting the adversary. This is the rule regardless of whether the civilians live in territory

---

[13] *ICRC Commentary*, p. 1458.

[14] *Id.*, p. 1458.

[15] *New Rules*, p. 340 (commenting on Article 54, Protocol I).

[16] *ICRC Commentary*, p. 1459 (footnote omitted).

controlled by that party or its adversary.[17] The *ICRC Commentary* notes:

> To deprive the civilian population of objects indispensable to its survival usually results in such a population moving elsewhere as it has no other recourse than to flee. Such movements are provoked by the use of starvation, which is in such cases equivalent to the use of force.[18]

Thus the counterinsurgency tactic of "draining the sea" -- or forcing civilians to move away from the guerrillas who live off them -- by means of depriving the civilians of food or rendering their fields useless for cultivation is prohibited by this article.

The article points out the most usual ways in which starvation is brought about but the list is not exhaustive. The words "attack, destroy, remove or render useless" are used to cover "all eventualities," including chemicals used to pollute water or defoliants used to destroy a forest, according to the *ICRC Commentary*.[19]

No less than chemicals or defoliants, use of contact land mines in agricultural areas or on paths to these fields has the effect of rendering the areas useless for food production, because no one will be able to plant there. The customary law prohibition against starvation of civilians as a method of combat forbids use of land mines to accomplish those ends.

Recording Requirement

The Land Mines Protocol contains a recording requirement in article 7(1)(a). The provision states that "[t]he parties to a conflict shall record the location of . . . all pre-planned mine fields laid by them."[20] Although the Land Mines Protocol does not define the term "pre-

---

[17] *ICRC Commentary*, p. 1459.

[18] *ICRC Commentary*, p. 1459.

[19] *ICRC Commentary*, p. 1458.

[20] Land Mines Protocol, Article 7(1)(a).

planned," an authority notes:

> Since 'pre-planned' means more than 'planned,' a 'pre-planned' minefield is, by its nature, one for which a detailed military plan exists considerably in advance of the proposed date of execution. Naturally, such a detailed military plan could not exist for the vast majority of minefields placed during wartime. In the heat of combat many minefields will be created to meet immediate battlefield contingencies with little 'planning' or 'pre-planning.'[21]

The provision for recording is designed to facilitate removal at the end of the conflict, primarily for the benefit of civilians. Thus, at the cessation of active hostilities, the parties are to "take all necessary and appropriate measures, including the use of such records, to protect civilians from the effects of minefields, mines, and booby traps."[22]

---

[21] Burrus Carnahan, "The Law of Mine Warfare: Protocol II to the United Nations Convention on Certain Conventional Weapons," 105 *Military Law Review* 73,82 (1984). The recording requirement applies only to the location of pre-planned minefields, not to the location of individual mines therein, nor to the composition or configuration of the mines within the field.

[22] Land Mines Protocol, Article 7 (3) (a).

CONCLUSIONS AND RECOMMENDATIONS

Conclusions

There is a very serious land mine problem throughout Angola, with certain parts of the country, such as Huambo, Bíe, Moxico and Cuando Cubango being particularly severely affected. All combatant forces, including FAPLA, FALA (the armed wing of UNITA), the Cubans and have been responsible for laying large numbers of land mines, especially anti-personnel land mines. A variety of countries, including the United States and Italy, have manufactured land mines that have been used in Angola, and will continue to be used in the present fighting.

Most of the mines have been laid without markings or warnings to the civilian population, and a large proportion have been laid in such a way that their victims are almost guaranteed to be civilians. As a result, a very minimum of 15,000, and probably more than 20,000 Angolans, are currently amputees, as a result of land mine accidents, and many thousands more have been killed. Even if a lasting peace is established, the human impact of the land mines is likely to increase in the short term, with the return home of refugees and displaced people and attempts by civilians to reclaim their villages, fields and pastures, and to travel along roads and paths.

Facilities for the evacuation, emergency treatment, hospital treatment and physical and social rehabilitation of land mine victims are inadequate and not improving. Hospital facilities are poor. More than 5,000 prostheses are required each year; current production is well under one third of that number. The social needs of land mine victims are not attended to adequately.

Should peace return, land mines will continue to present severe obstacles to the economic development of the country, the implementation of relief programs, and the return of refugees. Large areas of Angola will remain out of bounds for civilians until land mines are cleared and the community regains confidence in the land.

Recent and current initiatives to clear land mines are inadequate. Efforts by the Angolan government have suffered from lack of equipment, finance and coordination, and the international community does not appear to have made demining programs in Angola a high priority. Some major potential donors, such as the US, are almost completely absent from land mine eradication initiatives. This is a disgrace. The US continues to refuse to confirm whether it supplied land mines to UNITA, but whether it did or not, the extent of its military and political support to UNITA means that it bears a share of the responsibility for UNITA's abuse of land mines. Other countries, notably Great Britain, appear to be motivated primarily by commercial considerations. Ironically in view of its highly destructive military interventions in Angola, the South African clearance efforts appear to have been the best recent foreign initiative.

In theory, the use of land mines is subject to international law, namely the Land Mines Protocol of 1981.[1] Though applicable on its face only to international armed conflict, the main provisions of the Proctocol have the status of customary international law and, accordingly, apply to the conflict in Angola. Accordingly, international law prohibits the direct use of land mines against the civilian population, the "indiscriminate" use of these weapons, and their use in such a way as to cause civilian casualties out of proportion to the military objective envisaged. Mines are supposed to be marked, and the mining of built-up areas is prohibited unless the mines are placed in or close to a military target.

It is clear from this report that, in practice, all parties to the conflict in Angola have routinely abused these provisions. There appear to have been no serious or systematic attempts to minimize civilian casualties from mines. Indeed, the very types of anti-personnel land mine used and the strategies for their deployment suggest that the provisions of the Protocol would have been impossible to enforce without a major change of military strategy by all sides. It is hard to avoid the conclusion that one of the purposes of the random dissemination of these mines in inhabited areas was precisely to cause excessive civilian casualties and thereby terrorize the population. Anti-tank mines have proved less dangerous for civilians, but that is merely because of their design (specifically the fact that very heavy pressure is needed to initiate an explosion), not because the warring parties have heeded the provisions of international law. The experience of Angola suggests that the Land Mines Protocol has been wholly ineffective.

---

[1] The Protocol on the Prohibition and Restrictions on the Use of Mines, Booby Traps and Other Devices is part of the 1981 Convention on Prohibition or Restriction on the Use of Certain Conventional Weapons which may be Deemed to be Excessively Injurious and to have Indiscriminate Effects.

Recommendations

I.   **General**

The experience of Angola has shown that anti-personnel land mines present a serious and long-term threat to civilians, far in excess of any short-term military advantage that may be gained. Accordingly, we believe that the use of anti-personnel land mines should be banned altogether.

II.   **To the Angolan Government**

(1)   The Angolan government should take immediate steps to set up a systematic and coordinated mine clearance program that will eradicate mines from all areas that are used by civilians. Essential components of this program involve:

*   The training of surveyors, mine eradication teams, and team coordinators.

*   A survey of mine-infested areas and the drawing up of a systematic national mine-eradication plan.

*   Clearance of areas extending beyond major roads, railroads, bridges and built-up areas, such as minor roads and paths, the verges of roads, riverbanks, and other areas that local communities believe to be mined.

*   The clear marking of areas that remain mined, and informing the local community of the location of these mined areas.

*   The destruction of all land mines in situ or after lifting. No land mines should be stored.

(2)   The Angolan government should launch an international appeal to solicit funds and expertise for this program to be carried out.

(3) The Angolan government should sign and ratify the 1981 Land Mines Protocol and abide by its provisions in any future internal or international conflicts.

III. To FAPLA, FALA/UNITA, and the Cuban Armed Forces

(1) All combatant groups should provide expert personnel to assist demining efforts in Angola.

(2) All combatant groups should provide all available information to the demining commission about the types of mines they have used in Angola, the strategies of dissemination (including methods for preventing mine clearance) and the location of mines (including, wherever possible, minefield maps) to assist in clearance efforts.

(3) In the renewed fighting, all parties to the conflict should observe the provisions of the 1981 Land Mines Protocol under all circumstances.

IV. To the United Nations, Western Donors and Former Eastern Bloc Countries

(1) The United Nations should take the lead in coordinating with the Angolan government, drawing up a national mine eradication program, and soliciting funds and expertise.

(2) All countries that have provided land mines to Angola should contribute to the cost of the national mine eradication program.

(3) All countries that have manufactured land mines, or patented land mine designs that have been manufactured under license elsewhere, that have been used in Angola, should contribute to the cost of the national demining program.

(4) The United States is morally obliged to contribute to the national mine eradication program. Because the US has been one of UNITA's major backers, it bears a special responsibility for land mine eradication in Angola.